Guillermo César Mondragón Rodríguez

NOx-reductive Perovskite catalysts for lean-burn engines

Guillermo César Mondragón Rodríguez

NOx-reductive Perovskite catalysts for lean-burn engines

Development and characterization of NOx-reductive catalytic coating systems for lean-burn engines

Südwestdeutscher Verlag für Hochschulschriften

Impressum / Imprint
Bibliografische Information der Deutschen Nationalbibliothek: Die Deutsche Nationalbibliothek verzeichnet diese Publikation in der Deutschen Nationalbibliografie; detaillierte bibliografische Daten sind im Internet über http://dnb.d-nb.de abrufbar.
Alle in diesem Buch genannten Marken und Produktnamen unterliegen warenzeichen-, marken- oder patentrechtlichem Schutz bzw. sind Warenzeichen oder eingetragene Warenzeichen der jeweiligen Inhaber. Die Wiedergabe von Marken, Produktnamen, Gebrauchsnamen, Handelsnamen, Warenbezeichnungen u.s.w. in diesem Werk berechtigt auch ohne besondere Kennzeichnung nicht zu der Annahme, dass solche Namen im Sinne der Warenzeichen- und Markenschutzgesetzgebung als frei zu betrachten wären und daher von jedermann benutzt werden dürften.

Bibliographic information published by the Deutsche Nationalbibliothek: The Deutsche Nationalbibliothek lists this publication in the Deutsche Nationalbibliografie; detailed bibliographic data are available in the Internet at http://dnb.d-nb.de.
Any brand names and product names mentioned in this book are subject to trademark, brand or patent protection and are trademarks or registered trademarks of their respective holders. The use of brand names, product names, common names, trade names, product descriptions etc. even without a particular marking in this work is in no way to be construed to mean that such names may be regarded as unrestricted in respect of trademark and brand protection legislation and could thus be used by anyone.

Verlag / Publisher:
Südwestdeutscher Verlag für Hochschulschriften
ist ein Imprint der / is a trademark of
OmniScriptum GmbH & Co. KG
Heinrich-Böcking-Str. 6-8, 66121 Saarbrücken, Deutschland / Germany
Email: info@svh-verlag.de

Herstellung: siehe letzte Seite /
Printed at: see last page
ISBN: 978-3-8381-1182-7

Zugl. / Approved by: Bochum, RUB, Diss., 2009

Copyright © 2009 OmniScriptum GmbH & Co. KG
Alle Rechte vorbehalten. / All rights reserved. Saarbrücken 2009

Index

Acknowledgments		I
Abbreviations		II
Abstract		III

Chapter		Page
1.	Introduction	1
2.	Background and literature survey	5
	2.1 The state-of-the-art technologies for reduction of NO_x emission (post-combustion and after-treatments)	9
	2.2 Methods applied for NO_x-reduction under lean conditions	15
	2.3 Materials applied for NO_x-reduction under lean conditions	20
	2.3.1 Perovskites as catalysts for NO_x-reduction	24
	2.4 The EB-PVD coating method	27
	2.5 Open questions and material development required for NO_x-reduction under lean-burn conditions applying perovskites	29
3.	Motivation and objectives of the present investigation	31
4.	Materials and methods	33
	4.1 Synthesis of the catalysts	33
	4.1.1 The citrate route	34
	4.1.2 The co-precipitation method	35
	4.1.3 Manufacture of the integrated perovskite coating system	36
	4.2 Characterization methods	39
	4.2.1 Microstructure and composition analysis	39
	4.2.2 Crystal phase(s) analysis	41
	4.2.3 Surface area(s) measurements	41
	4.2.4 Determination of the catalytic activity	42
	4.2.4.1 Selective Catalytic Reduction of NO_x (SCR) with H_2	42
	4.2.4.2 Selective Catalytic Reduction of NO_x (SCR) with propene	43
5.	Results	47
	5.1 Microstructure and phase analysis of the powder catalysts	47
	5.1.1 Microstructure and crystallization path of Pd-substituted perovskite $LaFeCo_{(0.3)}$-Pd*	47
	5.1.1.1 Redox behaviour of the perovskite $LaFeCo_{(0.3)}$-Pd	54
	5.1.2 Microstructure and crystal phase(s) of Pd-substituted perovskites BaTi-Pd*	59
	5.1.2.1 Redox behaviour of the perovskite BaTi-Pd*	62
	5.1.3 XPS analysis of the La- & Ba- based perovskite catalysts	63
	5.2 Catalytic coating development	73
	5.2.1 Microstructure and phase analysis of the perovskite after the impregnation and calcination of the PYSZ coating	74
	5.2.2 Perovskite catalyst coated on top of monolithic cordierite substrate	81
	5.3 NO_x-reduction capability and N_2-selectivity of the powder catalyst	83
	5.3.1 NO_x-reduction with H_2 under excess oxygen	84

		5.3.2	NO_x-reduction with H_2 under lean conditions including H_2O and CO_2	86
		5.3.3	NO_x-reduction with H_2 under lean conditions in the presence of CO.......	90
		5.3.4	NO_x-reduction with C_3H_6 under lean conditions (HC-SCR)	92
		5.3.5	NO_x-reduction performance of the new catalytic coatings	97
6.	Discussion			99
	6.1	State of palladium in La- & Ba- based perovskites catalysts		100
	6.2	Catalytic activity of Pd-substituted perovskites towards the NO_x-reduction with H_2 under lean conditions		107
		6.2.1	Effect of the cobalt-content in the perovskite $LaFeCo_{(x)}$-Pd* on NO_x-reduction ....…..	107
		6.2.2	The role of cerium on the NO_x-conversion over the perovskite catalyst $LaCe_{(y)}Fe$-Pd* ....…...	113
		6.2.3	The catalytic performance of the catalyst BaTi-Pd* during H_2-SCR of NO_x ....…...............	117
		6.2.4	Catalytic activity of the catalysts BaTi-Pd and $LaFeCo_{(0.3)}$-Pd* for C_3H_6-SCR of NO_x ....…..	119
			6.2.4.1 The C_3H_6-SCR of NO_x activity of the catalysts Pd-$LaFeCo_{(0.35)}$ and $LaFeCo_{(0.3)}$-Pd* ....…................................	120
			6.2.4.2 The C_3H_6-SCR of NO_x activity of the catalysts Pd-BaTi and BaTi-Pd* ....…......................…........	121
	6.3	The Lanthanum perovskite coatings on top of EB-PVD PYSZ substrates		122
		6.3.1	Coating methods of the lanthanum based perovskite $LaFeCo_{(0.3)}$-Pd*	122
		6.3.2	Effect of the thermal ageing on the microstructure of $LaFeCo_{(0.3)}$-Pd-EB PVD PYSZ double coatings	123
		6.3.3	NO_x-reduction performance of the powder catalyst vs. the catalytic coatings	125
7.	Conclusions			129
8.	Outlook			135
	References			137
	Appendix			143

* Perovskite composition(s) in appendix A 4

I. Acknowledgments

The present work was made during my stay at the facilities of the High Temperature and Functional Coatings Division of DLR's (German Aerospace Center) Institute of Materials Research in Cologne Germany. I want to thank all the members of the Institute involved in someway during my PhD.

Thank to DLR and to Prof. Dr. Christoph Leyens (Head-Director of the Institute at that time when I came to Germany), Prof. Dr. Heinz Voggenreiter (Actual Head-Director of the Institute), and Prof. Dr. Stefan Reh (Managing Director of the Institute) for giving me the chance to do my PhD at the Institute of Materials Research.

This work would not be possible without the administrative support, scientific supervision, as well as critical review of the manuscript made by Dr. habil. Bilge Saruhan who made the scientific and technical supervision of my thesis. Thank you for your help.

Thank to Dr. Ing. Manfred Peters and Dr. Ing. Uwe Schulz for the financial and administrative support during this work and for giving me the facilities to publish and to present the results of this thesis in National and International Conferences.

Thank to Prof. Dr. Ing. Gunther Eggeler from the Institute of Materials and Prof. Dr. Wolfgang Grünert from the Laboratory of Industrial Chemistry of the University Bochum for their contribution to this work, interest to supervise and evaluate this thesis and for critical review of the manuscript.

I also thank SurMat (The International Max Planck Research School for Surface and Interface engineering in Advanced Materials) and Dr. Christoph Somsen for his help during the lectures at SurMat and administrative support at the Bochum University.

Special thank to all the technicians of the High Temperature and Functional Coatings Division for their help and support, K. Kröder, J. Brien, D. Peter and H. Mangers. Thank to R. Borath and U. Kreber for their technical support and introduction in the practical use of the scanning electronic microscope. Thank to K. Baumann for advice in the metallographic preparation of the samples for the SEM analysis, and Dr. M. Schmücker member of the Structural Ceramics and Functional Materials for his help and fruitful discussions. Thank to Dr. Klemens Kelm for the TEM analysis. I would like to thank Dr. B. Hildmann for the interesting discussions we had and for technical support in the XRD laboratory.

Thank to all my colleagues and co-workers for their support and help in solving the everyday problems involved with the experimental work, Dr. Ing. M. Stranzenbach, Dr. R. Ochrombel, Dipl. Ing. A. Ebach-Stahl. Thank to all the administrative personal that make possible all the administrative work during my stay at DLR, S. Schneider, C. Renn, Frau Patz, H. Frauenrath, F. Seidler and H. Schleicher.

Thank to Dr. J. P. Breen and his team of co-workers at Queen's University Belfast for their help and fruitful discussions during my stay at CenTACat which was financially supported by the EU Transnational Access Project which was funded by the EU under the "Structuring the European Research Area" specific programme Research Infrastructures Action (*RITA-CT-2006-025995*).

This work was financially supported by the Deutscher Akademischer Austauschdienst (DAAD) and the Mexican National Council for Science and Technology (CONACyT) through the fellowship 163638/200637.

Finally but not less important thank to my parents who supported me during my study, my brothers and sisters for their unconditional support, and to my girlfriend and beloved Heike Leifhelm for her patience and support during my work.

II. Abbreviations

A/F	Air to fuel ratio
BE	Binding Energy
BET	Brunauer Emmett Teller
CAEP	Committee on Aviation an Environmental Protection
CenTACat	Centre of the Theory and Application of Catalysis
DSC	Differential Scanning Calorimetry
EDS	X-ray Energy Dispersive Spectrometer
EPA	Environmental Protection Agency
EB-PVD	Electron Bean Physical Vapour Deposition
FWHM	Full Width at a Half Maximum
GHSV	Gas Hourly Space Velocity (h^{-1})
GSA	Geometric Specific Surface area (m^2/l monolith)
ICAO	International Civil Aviation Organisation
IR	Infrared
ISA	International Standard Atmosphere
JCPDS	Joint Committee on Powder Diffraction Standards
LTO	Landing Take-off Cycle
MCD	Multiple Concentration Detection
NTP	Non Thermal Plasma
NO_x	$NO + NO_2$
OBD	On board Diagnostic
PYSZ	Partial Stabilized Zirconium Oxide
PM	Particulate Mater
ppm	Parts per million
SCR	Selective Catalytic Reduction
SN	Smoke number
SLS	See level static conditions
SNCR	Selective Non-Catalytic Reduction
SNC	Storage NO_x Catalyst
SESAM	Sensor and Catalyst Test Unit
SEM	Scanning Electron Microscopy
SULEV	Super Ultra Low Emission Vehicles
TEM	Transmission Electron Microscope
TBC	Thermal Barrier Coating
TFR	Total Flow Rate
T_m	Melting point Temperature (°C)
T_s	Substrate Temperature (°C)
TWCs	Three Way Catalysts
UHC	Unburned hydrocarbons
VOCs	Volatile Organic Compounds
VIA	Vapour Incidence Angle
W/F	Weight of catalyst to flow ratio
XPS	X-ray Photoelectron Spectroscopy
XRD	X-ray Diffraction
λ	$= \dfrac{operated - air/fuel - ratio}{stoichiometric - air/fuel - ratio}$

III. Abstract

There is an urgent need for NO_x-reduction in exhausts from combustion engines. Perovskite structures are promising materials that can be applied to eliminate NO_x and other emissions under lean conditions. Perovskites allow the combination of various cations conserving its original crystal structure. For instance, precious metals substituted in perovskites yield a self healing principle over the nano-scaled segregation of noble-metal particles leading to improved ageing resistance and catalytic activity. In the present work, the development and the characterization of new perovskites doped with palladium for NO_x-reduction under lean conditions are reported. The combination of palladium and the perovskites synthesized here leaded to new catalytic sites that effectively reduce NO_x.

A new Ba-based perovskite, doped with palladium was synthesized by the co-precipitation method and calcined at 500°C - 900°C in order to analyze the crystallization path, development of specific surface area, and the morphological modifications and to correlate these parameters with the activity for H_2-SCR of NO_x. Propene and octane were also tested as reductants for NO_x-reduction under excess oxygen. Palladium was only partially incorporated into the Ba-based perovskite lattice as indicated by XRD and XPS. However, this perovskite was one of the best catalysts which displayed a maximum NO_x-conversion of 73 % and more than 80 % of N_2-selectivity during H_2-SCR of NO_x at 170°C.

La-based perovskites doped with palladium were synthesized by a modified citrate route. The crystal phase(s) of the perovskite(s) treated at different temperatures and under redox atmospheres (oxidizing and reducing conditions) were analyzed by means of DSC, XRD, SEM, XPS and TEM. The powder based catalysts were tested for H_2-SCR of NO_x. The effects of H_2O_{vapour} + CO_2 and CO in the feed to the NO_x-reduction and N_2-selectivity of the catalysts were also analyzed. A single phase catalyst was obtained by varying the elements at the B-site of the perovskite; these modifications were correlated with the catalytic activity for H_2-SCR of NO_x reactions. Similarly, the effect of the cerium content in the perovskite on the H_2-SCR of NO_x-reduction and N_2-selectivity was studied. According to this study the limit for cerium to form a solid solution in the lanthanum based perovskite lies below 0.05 fraction mol. Increase of cerium loading to 0.4 mol % caused the segregation of cerium oxide and improvement in the NO_x-conversion (72 %) and N_2-selectivity (73 %). Additionally the potential of the La- and Ba-based perovskites as SCR catalysts was tested employing propene as an alternative reducing agent. A selected La-based perovskite composition was coated on top of EB-PVD PYSZ and on cordierite substrates. The perovskite coating displayed good resistance against thermal shock loading at 500°C and 900°C after 100 cycles. The cordierite substrate coated with the perovskite (catalytic converter) was tested for C_3H_6-SCR of NO_x reactions under high space velocities and displayed 15 to 20 % of NO_x-conversion in the presence of 3.9 vol. % H_2O_{vapour}.

The present work contributes to the knowledge and understanding of physical-chemical surface and bulk properties of perovskite as catalyst for lean-burn applications. This work provides the state of the art in the H_2-SCR of NO_x technology and will surely serve as a reference material for further studies in this area. Moreover it contributes to understand the state of palladium in the perovskitic structures and provides the main steps for a successful transfer from the synthesis of the powder catalyst(s) to the prototype development as a catalytic converter.

Guillermo César Mondragón Rodríguez
Cologne Germany 2008

Chapter 1

1. Introduction

Nowadays, transport and mobility has becoming an inalienable right in industrialized countries. The everyday life is unthinkable without automobiles, trucks, airplanes, etc. The comfort of independent transportation is not for free and the by-products of the combustion process; NO_x, CO, UHC (Unburned Hydrocarbons), SO_x, CO_2, H_2O, which takes place in every engine are causing severe damages to the environment. Particularly, NO_x are produced mainly by reaction between oxygen with atmospheric nitrogen during fuel combustion. Fossil fuel combustion for energy production and processes such as the synthesis of adipic acid are other sources of NO_x. Environmental pollution due to NO_x-emission is a serious problem for environment and humans, since it causes the acid-rain and increases the ground ozone level.

Carbon dioxide emissions can be reduced if more efficient and light engines are produced. Fuel economy can be increased if gasoline engines work under lean conditions (i.e. lean burn engine). The oxidation of UHC and CO is relatively straightforward with an appropriate catalyst, but NO_x cannot be easily reduced under these conditions. For instance SO_2-emission can be significantly reduced, if natural gas is used as a combustion fuel. NO_x-emissions, however, are not avoidable solely by employing alternative fuels. Most of the gas emissions can be completely eliminated with the use of fuel cells. At the moment the use of the fuel cells alone is not the best cost effective solution due to the amounts of precious metal required. Combustion engines are still the most economic transportation systems but not the cleanest and that is why new and more effective catalysts are needed for the after treatment of the exhausts generated in every combustion process.

For vehicle manufacturers, there are two solutions; (a) an improved engine design to decrease NO_x formation and (b) the catalytic removal of NO_x-emission from the engine exhaust. NO_x can be efficiently removed from the exhausts of gasoline engines which work under stoichiometric air/fuel-mixture by using Three Way Catalysts (TWCs). However, the conventional three way catalysts do not eliminate NO_x under lean-burn conditions (excess oxygen). Among the advanced existing NO_x-removal methods for lean exhausts, the Selective Catalytic Reduction of NO_x using ammonia (NH_3-SCR) as reducing agent seems to be the most efficient procedure. However, an ammonia forming reducing agent (e.g. urea-water solution; Adblue) has to be carried along. Although NH_3-SCR can achieve 90% NO_x-reduction at steady-state conditions, side reactions that produce N_2O or NO occur. Moreover, a SCR system requires a system to monitor the dosage of reducing reagents injected into the exhaust. The amount of the non-reacted ammonia released during the SCR-process can be monitored and drastically reduced if the SCR systems are monitored. For this reason the vehicles in the future may have to be equipped with an automated on-road emission monitoring systems as expected to be ordered with the On-Board Diagnostic (OBD-III). The NH_3-SCR systems are usually equipped with an oxidation catalyst in order to reduce the amount of unreacted NH_3.

The use of hydrocarbons deliberately added to (or already present) in the feed has been attracting more attention as a method to eliminate NO_x under lean atmospheres in the last years. The use of methane as reducing agent would be probably the most economic way to eliminate NO_x since natural gas is abundant and available in most parts of the world. The issue here is to activate the methane molecules at the low exhaust temperatures. Indeed the NO_x-reduction values obtained with methane over platinum supported catalysts are too low for practical applications [1]. Higher hydrocarbons such as; propene, propane, butane, etc, provide higher NO_x-reduction levels under excess oxygen [2]. The challenge here is to selectively reduce NO_x under excess oxygen at a wide range of working temperatures (from 100° to 500°C). Under high oxygen partial pressures the hydrocarbons are oxidized instead of react with NO_x. The second issue is the selectivity of the catalyst under the mentioned reaction conditions. Over platinum supported catalysts a major part of the reduced NO_x are converted into N_2O [1]. In general the palladium based catalysts are slightly less active during NO_x-reduction but more selective in terms of the nitrogen production. Only a few studies of the HC-SCR of NO_x over palladium containing perovskites can be found in the literature [3]. This is a new research area and more studies, especially under reaction conditions that are close to the real applications, are urgently needed.

Hydrogen has been also applied as a reducing agent to eliminate NO_x under lean exhausts at temperatures below 300°C in model systems [4]. Platinum supported catalysts (e.g. Pt/SiO_2, Pt/Al_2O_3) display a very good activity for the H_2-SCR of NO_x below 200°C. The main problem here is the production of high volumes of N_2O as a side product [4]. The work of Costa et al [5, 6] have shown that platinum supported on perovskite based structures (e.g. $Pt/La_{0.7}Sr_{0.2}Ce_{0.1}FeO_3$, $Pt/La_{0.5}Ce_{0.5}MnO_3$) provides excellent selectivities to nitrogen over a relatively wide temperature window. These catalysts show very good resistance and stability even after the exposure to water vapour and 25-40 ppm of SO_2. Even though the platinum supported on perovskitic structures show excellent catalytic properties, it is necessary to reduce the amount of platinum or to completely replace this metal by cheaper alternatives such as palladium.

A challenging task would be to eliminate N_2O under lean conditions. Under certain reaction conditions (in the absence of oxygen) N_2O is easily decomposed into nitrogen and oxygen over $Fe_xCe_{1-x}O_2$ mixed oxides [7]. On the other hand the work of Yan et al [8] have shown that N_2O can be decomposed even in the presence of excess oxygen and water vapour in the stream on $M_xCo_{1-x}Co_2O_4$ (M = Ni, Mg) catalysts.

Modern TWC's catalytic converters are the preferred technology for emissions after treatment in gasoline engines. The requirements and capabilities of the converter determine its location in the exhaust flow. The thermal shock resistance, thermal durability, mechanical integrity, warm-up and low pressure drop, of the converter continue to be improved at more critical levels and for a longer duration [9]. The catalyst support must be able to withstand the different temperatures under driving conditions that range from room temperature up to 900°C depending on the position of the catalyst in the car. Extreme temperatures, which usually occur due to engine ignition problems such as misfiring, can cause the temperatures to exceed 1000°C. Under these extreme operating conditions the typical supported metal catalysts are deactivated resulting in a drastic efficiency decrease of the catalytic converter.

Catalysts with Perovskite structure (ABO_3) show interesting physical and chemical properties, such as a high catalytic activity for oxidation and redox reactions. Many studies on perovskites have shown that the direct decomposition of NO is strongly dependent on the oxidation state of the elements at the B-site, on the presence of highly active sites, on the catalyst surface area and porosity of the compounds [10]. Precious metals (i.e. platinum, palladium, rhodium, and ruthenium) can be stabilized into the lattice of perovskites. According to the publication of Tanaka et al [11], the proportion of the metal substitution depends on the perovskite composition and on the metal itself. The precious metal substitution in a perovskite based catalyst yields a special self-healing principle over the nano-scaled segregation of the noble-metal particles, which promotes the ageing-resistance at the long-term use of the catalyst.

Those precious metal-ions (e.g. Pd-cations) incorporated in perovskite structure can move out of the perovskite lattice (e.g. segregate as Pd-nano-particles) by a reversible redox-reaction in H_2 containing atmospheres. As a consequence, the thermally activated growth of the Pd particles which may otherwise occur during the long-term use of the catalyst, will be suppressed [12]. A synergistic effect will be achieved between the perovskite lattice and the metal(s) dissolved into the crystal lattice (i.e. palladium) leading to an improvement of the catalytic activity and N_2-selectivity of the catalyst. The interaction perovskite/precious metal can synergistically work together, modifying the redox properties and the selectivity of the catalyst.

In the present work the lanthanum and barium based perovskites were synthesized, characterized and tested for H_2-SCR of NO_x and C_3H_6-SCR of NO_x reactions. A modified citrate route was employed to synthesize the lanthanum group of perovskites. The co-precipitation method was applied to prepare the barium based oxides. Powder catalysts were used for phase characterization by means of XRD and TEM. Surface chemistry and the state of palladium in or on the perovskites were analyzed with XPS. The Pd-substituted (here is meant to the incorporation of palladium into the crystal lattice of the perovskite) and Pd-supported (prepared by direct impregnation) perovskites were reduced in hydrogen containing atmospheres in order to prove the reversible or non-reversible diffusion of palladium and finally analyzed with XRD and XPS. The H_2-SCR of NO_x reactions were carried out at CenTACat Queen's University Belfast. The C_3H_6-SCR of NO_x reactions were carried out at the facilities of the High Temperature and Functional Coatings division of DLR's Institute of Materials Research.

A selected perovskite composition was coated on top of the EB-PVD PYSZ coatings and on cordierite substrates. The idea is to demonstrate the applicability of this perovskites as catalysts to treat the automotive exhausts or eventually for applications in gas turbines. The lanthanum perovskite coated on the cordierite substrates was tested for C_3H_6-SCR of NO_x under reaction conditions that are close to the real conditions (to simulate automotive engine exhausts). The presence of water and increase of the oxygen concentrations in the feed during the C_3H_6-SCR of NO_x on the catalytic coatings were studied. The test of the catalytic converters is seldom reported in the literature [13]. The challenge in this case is to test prototypes in the laboratory under reaction conditions present during real driving conditions.

Chapter 2

2. Background and literature survey

The combustion of hydrocarbons is carried out to satisfy much of the actual energetic requirements, however, produce pollutants that are discharge to the atmosphere. Among these contaminants are the nitrogen oxides (NO_x) which are the source of severe environmental problems such as acid rain, smog formation, global warming and ozone layer depletion. Acid rain which causes significant damage to forests and lakes is created when sulphur and various nitrogen oxides NO_x combine with atmospheric moisture. Acid rain can contaminate drinking water, damage vegetation and aquatic life, and erode buildings and monuments. Acid rain can also cause corrosion to airplanes, cars, and boots. Fig. 2.1 shows the atmospheric chemistry of the photochemical smog and the mechanism formation of acid rain. The principal oxides, the components of the polluting emission gases formed by the combustion of biomass and fossil fuels are mainly the nitric oxide (NO), the nitrogen dioxide (NO_2) and the dinitrogen monoxides (N_2O) which are collectively called NO_x. NO is the primary form in the combustion products (typically 95% of total NO_x) [14]. NO_x is produced not only by burning of fuels and biomass but also by lightening, the oxidation of NH_3 (produced by microbial decomposition of proteins in the soil), and volcanic activity. N_2O is mainly produced in the nature by microbial activity. The human contribution to the release of N_2O to the atmosphere is about 30-40 % of the total N_2O emissions. This includes the activities such as the adipic acid production, the nitric acid manufacture, the fossil fuels and biomass combustion, and the use of fertilizers. N_2O is converted into N_2 and NO in the stratosphere and it contributes to the destruction of the earth's protective ozone layer and to the green house effect [15].

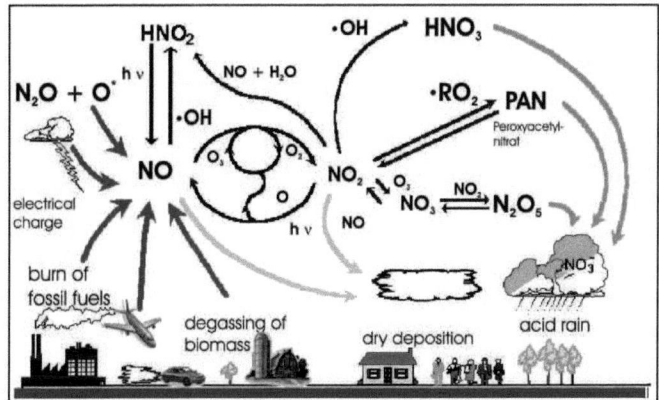

Figure 2.1 NO_x emissions and photochemical smog formation.

In the past years a lot of efforts have been made in the basic and applied research oriented to find out solutions and alternatives to reduce NO_x and other contaminant gas emissions. In the last 30 to 50 years an intensive work has been made to develop the appropriate technology to control and eliminate NO_x, CO, unburned hydrocarbons, soot, between others. For the after treatment of rich exhaust emissions in gasoline engines, the Three-Way-Catalyst (TWC) represents the most important technology. For diesel engines in trucks and industry (lean exhausts) the Selective Catalytic Reduction (SCR) with ammonia, urea as a source of ammonia, or even with methane or higher hydrocarbons represent the best technological approach at the moment. NO_x-traps or adsorbents combined with processes like plasma or microwaves have been also designed to activate the catalyst surface for a more effective and the long term reduction of the emissions.

Technologies to prevent the formation of NO_x such as the catalytic combustion have been also investigated. This concept focused to decrease the working temperatures during combustion and thus avoiding the NO_x formation (Thermal NO_x) mainly in gas turbines. Although the variety of options to control and reduce the exhausts emissions there are still many material issues to be overcome. The actual technologies are in some cases not precise enough or simply not fully developed for practical applications. So, it will be difficult to meet the forthcoming emission regulations. Development of new concepts and control systems must be made to avoid the production of contaminant gases and to promote the elimination of toxic emissions. Improved efficiency in fuel consumption will also contribute to the reduction of contaminants.

The catalytic methods are probably the most used and adequate, because they transform NO_x into nitrogen, and thus represent a true solution to the pollution problem. On the other side, the formation of the by-products such as N_2O with the use of new catalysts is extremely important and must be seriously taken into consideration. The detailed description of the functioning principle of the methods applied to eliminate NO_x and other emissions are discussed in sections 2.1 and 2.2.

Emission regulations in automotive and aircraft engines

Automotive and aircraft engines produce an enormous amount of toxic emissions as by-products during combustion. For this reason the European Union has released stricter emission regulations for both gasoline and diesel cars. One of many legislative instruments to reduce and control the amount of toxic emissions is the Gothenburg Protocol adopted on 30 November 1999. The protocol sets emission ceilings for 2010 for four pollutants: sulphur, NO_x, VOCs (Volatile Organic Compounds) and ammonia. Once the protocol is fully implemented, Europe's sulphur emissions should be cut by at least 63%, its NO_x emissions by 41%, VOC emissions by 40% and ammonia emissions by 17% compared to 1990. The protocol also sets tight limit values for specific emission sources (e.g. electricity production, dry cleaning, cars) and requires the best available techniques to be used to keep the emissions down. It has been estimated that once the Protocol is implemented, the area in Europe with excessive levels of acidification will shrink from 93 million hectares in 1990 to 15 million hectares. The number with excessive ozone levels will be halved. Consequently, it is estimated that life-years lost as a result of the chronic effects of ozone exposure will be about 2,300,000 lower in 2010 than in 1990, and there will be approximately 47,500 fewer premature deaths resulting from ozone and particulate matter (PM) in the air. The exposure of the vegetation to excessive ozone levels will be 44% down than in1990.

In 1992, the first emission standards regulation for controlling the release of the nitrogen oxides and the particulate matter was introduced in Europe. This legislative instrument was called "EURO I" and since then a series of stricter emission standards regulations have been approved and set into practice. The EURO IV is the actual set of emission standards applied to all the new on road vehicles from 2005 or later. It limits the diesel car emissions to 0.25 g/km of NO_x and 0.025 g/km of Particulate Matter (see Table 2.1), the petrol cars to 0.08 g/km of NO_x (see Table 2.2) and the Heavy Goods Vehicles to 3.5 g/kWh of NO_x and 0.02 g/kWh of PM. For heavy duty diesel trucks Euro IV will be succeeded by EURO V, to be introduced in 2008 and coming in force in 2009. For passenger cars, a EURO V standard is not set yet but is currently considered by the European Commission in the proposed regulation COM (2005) 683 final [16]. The EURO VI will enter into force by September 2014.

Table 2.1 Emission Standards in Europe for diesel cars.

	CO (mg/km)	Particulate Matter (mg/km)	NO_x (mg/km)
EURO 3 (2000-31.12.2004)	640	50	500
EURO 4 (from 1.1.2005)	500	25	250
EURO 5 Proposal	500	5	200

Table 2.2 Emission Standards in Europe for gasoline cars.

	CO (g/km)	HC (mg/km)	NO_x (mg/km)
EURO 3 (2000-2005)	2,30	200	150
EURO 4 (from 2005)	1,0	100	80
EURO 5 Draft proposal	1,0	75	60

Emission standards for aircraft engines

Under ideal conditions, the combustion of kerosene-type fuels produces carbon dioxide (CO_2) and water vapour (H_2O), the proportions of which depend on the specific fuel carbon to hydrogen ratio. At cruise conditions the combustion products constitute only about 8.5 % of the total mass flow emerging from the engine. Of these combustion products, a very small volume (about 0.4 %) of residual products arise from non-ideal combustion processes (soot, HC, and CO) and the oxidation of atmospheric nitrogen to the NO_x formation. Table 2.3 gives typical emission levels for various operating regimes for an engine type TF (Turbo fan) with the identification number 7GE099. The emission values are quoted as an emission index in units of grams of emitting species per kilogram of the fuel burned. The Table 2.3 shows that the maximal production of the NO_x emissions takes place at the first two stages of the flight cycle (at high power settings). These emissions are directly related to the fuel consumption of the engine in its various flight phases. The NO_x emissions, as CO, HC, and soot are strongly influenced by a wide range of variables but particularly the engine power setting and the ambient engine inlet conditions. CO and HC are products of an incomplete combustion. They are highest at low power settings, when the temperature of the air is relatively low and the fuel atomization and the mixing processes least efficient. This problem area is proven to be responsive to improvements linked to detailed studies of the basic fuel/air mixing processes. The majority of NO_x (thermal NO_x) are generated in the highest temperature regions of the combustor-usually in the primary combustion zone, before the products are diluted. The combustion zone temperature depends on the combustor inlet air temperatures and pressure, as well as the fuel/air mass ratio.

Table 2.3 Typical emission index (g/kg) levels for the engine operating regimes.

Operation mode	Power set %	Time (min)	Fuel flow (kg/s)	Emission Indices (g/kg)			
				HC	CO	NO_x	Smoke number
Take-off	100	0,7	4,69	0,04	0,08	50,34	4,1
Climb out	85	2,2	3,67	0,03	0,07	35,98	2,5
Approach	30	4,0	1,13	0,06	1,98	16,5	1,45
Idle	7	26,0	0,38	4,24	39,11	5,19	0,87
LTO Total fuel (Kg) or emissions (g)*			1546	2559	23858	34888	----

*ICAO Engine Exhaust Emissions Data bank
LTO Landing Take Off.

To control pollutants from aircraft in the vicinity of airports, the ICAO (international Civil Aviation Organization) established the emissions measurement procedures and the compliance standards for soot (measured as smoke number-SN), the unburned hydrocarbons, carbon monoxide, and the oxides of nitrogen. A landing take-off cycle (LTO) was defined to characterize the operational conditions of an aircraft engine within the environs of an airport; this LTO cycle is illustrated in Fig. 2.2. The standards are applied to all newly manufactured turbojet and turbofan engines that exceed the 26 kN rated thrust output at International Standard Atmosphere (ISA) sea level static conditions (SLS). The smoke standards took effect in 1983, and those for gaseous emissions took effect in 1986. The measurements of the exhaust emissions of a single engine are performed at the manufacturers test facilities as part of the certification process. The engine emissions are given for a standardized LTO cycle represented by an engine power setting of 7 (taxiing), 30 (approach), 85 (climb-out), and 100% (take-off) of rated output and the given times in mode. Together with the fuel flow, the emission indices of HC, CO, and NO_x in g/kg of fuel burned and maximum SN are reported. The emissions measurements are taken at the exit plane of the engines exhaust nozzle (within 0.5 nozzle diameter). The kerosene-type fuel complying with specified properties-density, heat value, boiling points, aromatics (15-23 % volume), sulphur (less 0.3 % mass), hydrogen (13.4-14.1 % mass) is used. No additives for a smoke suppression are allowed. For regulatory purposes, therefore, a statistically based correction is used to account for an engine-to-engine variability resulting from manufacturing tolerances.

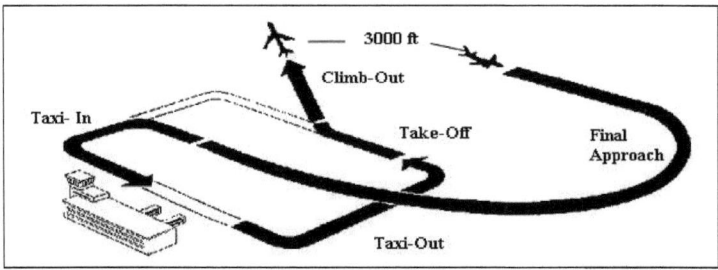

Figure 2.2 ICAO landing and take-off cycle (LTO).

Aircraft are required to meet the engine certification standards adopted by the council of ICAO. The standard for NO_x, carbon monoxide, unburned hydrocarbons, smoke and vented fuel emissions was first adopted in 1981, and then made more stringent in 1993, when the ICAO reduced the permitted levels by 20 % for newly certificated engines, with a production cut-off on 31 December 1999. In 1999, the council further tightened the standards by about 16 % on average for engines newly certificated from 31 December 2003. In February 2004, the Committee on Aviation an Environmental Protection CAEP/6 (six meeting of CAEP) agreed to establish more stringent international consensus emission standards for newly certified aircraft engines (implementation date, January 2008) [17]. The CAEP/6 NO_x standards generally represent about a 12 % increase in stringency from the standards promulgated in the CAEP/4 NO_x standards. Recent standard emission indices for different engines can be found on the Civil Aviation Authority web page [18]. Internationally, substantial engine research is in progress, with goals to reduce the Landing and Take- off cycle (LTO) emissions of NO_x by up to 70 % from the today's regulatory standards, while also improving the engine fuel consumption by 8 to 10 %, over the most recently produced engines, by about 2010.

2.1 The state of the art technologies for reduction of NO_x emission (post- and after treatments)

The early contribution to the development of the catalytic converters was made at the early beginning of the 19th century by a French chemist, Michel Frenkel with his patent describing the "deodorizing" function of exhausts using air blow in a fan was firstly reported [19]. The first catalytic converter, which was an oxidation catalyst, was invented and patented in 1962 by the engineer Eugene Houdry [9]. Three years later the Engelhard Company patented a catalytic converter based on the platinum group materials. The modern catalytic converter, based on platinum group metals deposited on a ceramic honeycomb, was industrially used in the US and Japanese Markets after 1974. At the beginning the purpose of using a catalyst was to oxidize carbon monoxide (CO) and unburned hydrocarbons (UHC) to produce only CO_2, denominated as a Two Way Catalyst (TWC). The ceramic "honeycomb" made of cordierite ($2MgO.2Al_2O_3.5SiO_2$) was used as support since the late 1970s. The honeycomb arrangement allowed incorporating up to 400 cells per cm^2 providing a relatively high contact surface areas. Ceramic substrates with higher cell densities up to 900 cells per cm^2 were developed in the following years. Simultaneously the fine steel foils (Fecralloy) appeared and still are applied as a support for the active phase providing a major cell density with a very low pressure drop and the additional advantages of a high thermal resistance and a high melting point (see Fig. 2.3).

Figure 2.3 Metallic substrate employed for automotive catalytic converters, courtesy of UMICORE®

At the early 1980s increasing concern about the emission of toxic gases such as soot and photochemical oxidants including ozone, nitrogen oxides (NO_x), sulfur dioxides (SO_x), promote the establishment of new legislation rules to control these pollutants mainly in the LA California area in the United States of America. In this year the registered vehicles surpassed 17 million and the vehicle miles traveled were about 155 billion, giving cumulative LA California vehicle emissions for NO_x and HC of 1.6 million tons/year. Only the precious metals such as Pd, Rh, and Pt were able to remove the pollutants in the short residence time determined by the large volumetric flows of the exhaust in relation to the size of the catalyst. Initially only Pd and Pt dispersed in a high surface area "wash-coat" were used in various proportions as oxidation catalysts. The use of Rh in the catalytic converters came into action with the development of an innovative catalytic system called the Three Way Catalyst. Initially Pt and Pd were used as oxidation catalysts, Rh was introduced with the introduction of the Three Way Catalyst for the catalytic reduction of the nitrogen oxides [20]. The amount and presence of the noble metal(s) used in the catalytic converters has been determined by many factors; the chemical and thermal resistance of the catalyst, the technical difficulties concerning the exhaust stream composition, its temperature and velocity, and the high prices of Rh and Pt due to the increasing demand driven by the industrial production of automotive catalytic converters in the US, Japan and Europe.

The Three Way Catalyst (TWC)

The Three Way Catalyst (TWC) consisted of basically a) an oxidation unit, usually Pd and/or Pt supported metal on ceramic or metallic substrate, b) a reduction catalyst, usually Rh supported catalysts and c) the oxygen sensor. Nowadays the modern Three Way Catalysts operate in a close loop system including a lambda- or oxygen sensor to regulate the air-fuel ratio. A novel concept of the TWC implies the oxidation and the reduction catalyst in only one component equipped with a gas oxygen sensor. The TWCs can simultaneously eliminate CO, UHC and NO_x; CO and HC are converted under oxidizing conditions to CO_2 (i.e. with excess air or $\lambda>1$) while NO_x are reduced to N_2, requiring excess fuel. The reactions have the highest efficiency in a small air-to-fuel ratio (A/F=14.6) window around the stoichiometric point, when air and fuel are in chemical balance $\lambda=1$. Fig. 2.4 shows clearly the highest NO_x conversion near the stoichiometric point, at higher air/fuel ratios (A/F=14.6) the NO_x conversion drops down, which limits to some extend the increase in the fuel efficiency. A decisive advance in the emission control has been made with the improvement in electronics over the last 30 years. The development of new catalysts coupled with the O_2-gas sensors have narrowed the air-fuel ratio window allowing to reach higher fuel efficiency and less NO_x. However, further research is needed to optimize the functioning of the engine control system and reduce the emissions to a minimum.

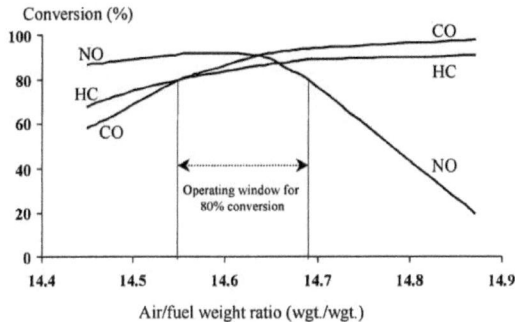

Figure 2.4 Plot of the conversion efficiency for the three species as a function of the mass air-fuel ratio (λ value).

From the technological point of view the most likely strategy to improve the fuel efficiency in mobile units is the development of light-structural components (to reduce weight). Other methods include modifications in the combustion stage of the aircraft and the engine technology from automobiles, trucks and other vehicles to further improvements in the fuel consume [21]. In gasoline engines the fuel efficiency is coupled with the oxygen consume during the fuel combustion. Use of an excess air would result in a better fuel economy at expense of higher NO_x-emissions. The nitrogen oxide can be effectively eliminated by the TWC but only under rich conditions (see Fig. 2.4).

A partial lean-burn strategy, where NO_x are adsorbed under a short lean conditions period and then reduced to N_2 under a rich phase has been proposed as a practical approach to eliminate NO_x in rich streams. Fig. 2.5 show a schematic representation of the arrangement of the TWC components combined with a NO_x storage unit and an oxidation catalyst. Under rich conditions almost hundred percent of NO_x, CO_x and UHC are eliminated. The oxygen sensors help to maintain the engine working near the stoichiometric point (λ=1). Delay between the sensor measurement and the working conditions of the engine causes a delay that result into cyclic lean periods. Under lean-burn conditions the TWC do not reduce NO_x. The storage of nitrogen oxides is carried out under the excess of oxygen while UHC and CO_x are oxidized. Under a lean period the NO_x-trap is regenerated. Issues related to the deactivation of the NO_x-trap, usually barium based materials, still remains as a research opportunity. Low sulfur fuels would help to solve the deactivation of the NO_x-traps caused by the surface sulfate formation. The NO_x-trap catalysts are discussed in more detail in section 2.2.

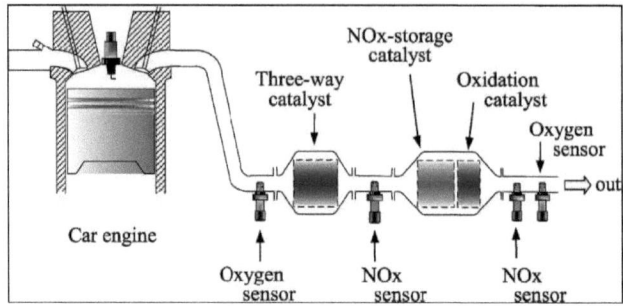

Figure 2.5 Partial lean-burn strategy to eliminate NO_x, CO_x and UHC from gasoline engines [22].

An alternative approach to reduce NO_x under practical applications (i.e. gasoline engines) is by using the unburned hydrocarbons already present in the exhaust. The amount and type of the hydrocarbons depend of course of the fuel quality and the working conditions of the engine. Table 2.4 shows typical concentrations of exhaust gas constituents for a gasoline engine after the FTP [23]. Important is that the oxygen concentration in the exhaust is relatively low (0.51 vol. %) known as rich conditions. Under these conditions most of the NO_x react with the hydrocarbons present in the exhaust. An increase in the oxygen concentration (lean conditions) would lead to a better fuel efficiency (lean-burn engine). One of the main issues is to selectively eliminate NO_x with the reducing agents in the exhaust in the presence of excess oxygen. Under oxidizing conditions the hydrocarbons are selectively oxidized, particularly at relatively high temperatures (i.e. > 250°C depending on the reaction conditions) and NO is also oxidized to form NO_2 depending on the precious metal [24]. The lean-burn engines cannot meet the strict NO_x emission regulations and need additional alternatives to reduce NO_x.

Table 2.4 Concentrations of exhaust gases (average emissions over an FTP [b] test) [23].

Component	Concentration
HC^a	750 ppm
NO_x	1050 ppm
CO	0.68 vol. %
H_2	0.23 vol. %
CO_2	13.5 vol. %
O_2	0.51 vol. %
H_2O	12.5 vol. %

[a] Based on C3
[b] Federal Test Procedure

NO_x elimination from exhausts in diesel engines

Diesel engines are more efficient in terms of the fuel consumption compared to gasoline engines. The combustion of diesel requires high air volumes which produces emissions with high oxygen concentrations. Under strong lean conditions the oxidation of the pollutants such as; CO and UHC is relatively straightforward, but the NO_x elimination becomes difficult. In addition to this problem exhaust temperatures of the diesel engines are around 100 to 150°C lower compared to the exhausts from gasoline engines under real driving modes [9]. In diesel engines the exhaust temperatures vary from ca. 120°C at a speed of 30 Km/h to almost 600°C at 120 Km/h. On the other hand due to the nature of the diesel combustion process, high amounts of particulate matter (PM) are formed. In recent years, improvements in engine control and in combustion engineering have been recorded resulting in increased fuel efficiency. However, the increased efficiency in fuel consumption brings about the formation of extremely fine particulates that eventually reach the environment. There is increased concern about the health effects of very small soot particles.

Particularly complicated is the selective removal of NO_x from exhausts with high oxygen content. A practical way to get rid of nitrogen oxides would be by letting NO_x to react with unburned hydrocarbons (UHC) in the exhaust stream. That is why many research efforts are oriented to find out catalysts that selectively reduce NO_x with hydrocarbons in the engine exhaust stream (i.e. CH_4, C_3H_6, C_4H_{10}, C_8H_{18}, etc). An alternative strategy is to add different hydrocarbons or even hydrogen in the exhaust gases to reduce NO_x, like in the case of the NH_3-SCR technology where NH_3 or Urea as a source of ammonia, are sprayed in the exhaust. NH_3 and Urea-SCR concepts are currently applied to eliminate NO_x from stationary power plants, nitric acid plants, coal and oil fired boilers, process heaters and chemical plants, around 90% of NO_x conversion can be achieved. The main groups of commercial catalysts used to remove NO_x with NH_3 can be classified into V_2O_5/TiO_2 mixed oxides, zeolites and Platinum based catalysts [25]. The major obstacle of V_2O_5/TiO_2 mixed oxides is due to the formation of SO_3, presumably because of the presence of high V_2O_5 contents. Zeolites are poisoned under the presence of water and thermally deactivated at higher temperatures (345-590°C). Pt based catalysts, which are mainly used for the low temperature applications (150-300°C) produce large amounts of N_2O in addition to N_2.

For practical applications in diesel engines an advanced dosing system for the reducing agent is necessary to provide the right amount of reductant during all driving conditions. Malfunction of the dosing systems lead to the release of unreacted ammonia in the exhaust, this is called "NH_3-slip". Oxidation catalysts are usually placed downstream to transform the excess ammonia into NO. In mobile units an additional tank has to be carried on to provide the reducing agent bringing about the problem of refilling at the right time and place. Even though the drawbacks of the actual NH_3 or Urea-SCR technology (commercially known as Adblue), the experience have shown that the use of urea in form of 32.5 vol. % aqueous solution is the best available method to reach the emission levels set for diesel engines. Moreover NH_3 emission will be regulated in the forthcoming EURO norms. Thus, new improvements and innovative catalyst development are required in order to reach the future emission regulations.

In Summary, the removal of NO_x from exhausts in the presence of excess oxygen will play a key role in the next years. Diesel engines are becoming more popular, particularly in Europe. It can be foreseen that the number of diesel engines on the roads will increase in the near future. NO_x elimination at the whole range of the exhausts temperatures from diesel and lean gasoline engines (100-600°C) must be covered. A challenging problem is to reach the catalyst working temperature as quickly as possible in order to reduce the "Cold start emissions" (50 – 100°C). During the cold start period most of the UHC and NO_x are emitted. The so-called cold start phase takes 100 to 200 seconds before the catalyst warms up. Different approaches and catalysts applied to eliminate NO_x at low temperatures (<200°C) are discussed later on.

The catalytic combustion method

At the moment the main aspects to improve in aircraft and gas turbines are; the NO_x emission reduction, the fuel efficiency and the noise decrease. There are concrete goals to reduce emissions of NO_x by up to 70 % from today's regulatory standards for aircraft engines, while improving the engine fuel consumption by 8 to 10 %, by about 2010 [17]. A 20 % improvement in fuel efficiency is projected by 2015 and a 40 to 50 % improvement by 2050 relative to aircraft produced today [26]. Engine efficiency improvements reduce the specific fuel consumption and most types of emissions; however, contrails may increase and, without advances in the combustion technology, NO_x-emissions may also increase. Two options are considered in order to reach these values. The first method to improve the fuel consumption is to reduce weight. The second option is the increase of the temperature in the combustion chamber. Disadvantage of the later is the increasing concentration of the thermal NO_x formed at temperatures higher than 827°C [27].

Thermal NO_x are formed in every combustion process by the reaction between N_2 and O_2. The formation of the thermal NO_x is rapid in the high-temperature lean zones of the flames. The mechanism of the NO-formation with nitrogen and oxygen as reactants in combustion processes is known since 1946 [28]. The rate of NO formation increases exponentially with temperature. The catalytic combustion is a method employed to reduce the combustion temperatures and thus the formation of the nitrogen oxides [28]. By employing a catalyst in the combustion stage of the gas turbines the combustion temperature and NO_x can be reduced. This is a concept often reported in the literature as a feasible method to reduce the nitrogen oxides formation without cost penalties [29].

The first catalytic combustor may have been made in France 1916 as reported by H. Sadamori [30]. Nowadays, design and material development in the new catalytic combustors is broadly reported in the literature [31, 32]. A schematic construction of a catalytic combustor designed by the group of Sadamori in 1995 is shown in Fig. 2.6. The ceramic combustion catalyst (ceramic body + catalyst) was exclusively made of manganese doped hexaaluminates which showed no coating cracking, however, the thermal shock resistance and the mechanical strength were low because of the high thermal expansion ratio of this material. This catalytic combustor was designed to be applied in gas turbines having a capacity of 1000 to 1500 kW, a maximum inlet temperature of 1100°C and 10 atm of absolute pressure. Less than 40 ppm of NO_x emission was set as a target to be reached using natural gas as a fuel.

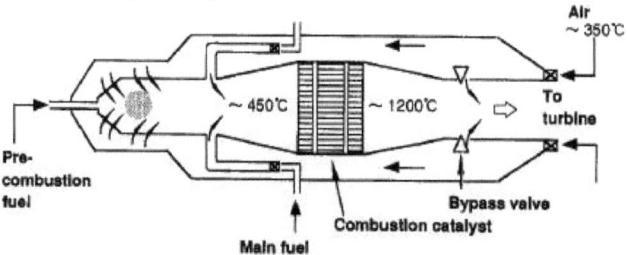

Figure 2.6 Design of a catalytic combustor proposed by Sadamori et al [33].

A newer catalytic combustor design has been suggested by the research group of Carroni et al in 2002 [34]. Fig. 2.7 displays a schematic diagram of a catalytic combustor design by Carroni et al [34]. In this case, the catalyst is placed at the pre-combustion stage of the gas turbines at the different temperatures found in modern gas turbines. In his study Carroni prepared a palladium supported catalyst (3-7 wt. % Pd) on top of a cordierite honeycomb structure (200-400 cpsi) and a Fe-Cr monolithic alloy (0.05 mm sheet thickness and ~107 cpsi). The catalytic combustors in this case was applied at high gas space velocities (620 000 – 1 080 000h^{-1}), at a catalyst inlet temperature of 450°C and pressures up to 30 bar. It was targeted to reduce NO_x < 3 ppm and CO < 10 ppm in the presence of 15 vol. % O_2 at 50 – 100 % load, using natural gas. The catalytic combustor design comprises two stages, a highly reactive catalyst for the low temperature conversion of CH_4 (350°C < T_{in} < 450°C), and a second catalyst that converts the fuel at higher temperatures (T_{in} approx. 700°C, T_{out} > 900°C).

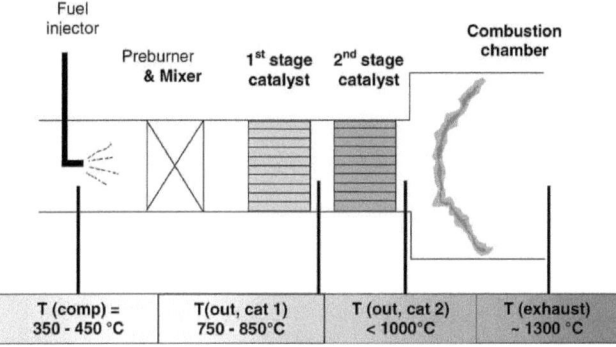

Figure 2.7 Principle of the catalytically stabilized combustion (CST), Carroni et al [34].

Although the system works functionally, the material related issues regarding the durability of the catalyst in the catalytic combustion technology remain to be solved. Palladium supported catalyst, which are the most suitable for the first stage suffer from ageing, sintering and poisoning under the extreme temperatures. Catalytic combustion is an alternative technology under development that has the potential of reaching low emission levels and cost reduction [35]. However in order to test prototypes under practicable conditions extremely high flow rates, high pressures and temperatures are needed. Under laboratory conditions it is easier to reproduce the conditions found in automotive exhausts i.e. the low stream temperatures (<600°C) and the gas concentrations at atmospheric pressure.

2.2 Methods applied for NO_x-reduction under lean conditions

The Selective Non-Catalytic Reduction (SNCR)

The elimination of NO_x (NO + NO_2) from exhaust gases which contains excess oxygen (lean-burn conditions) is a challenging issue. Up to know, for conventional power plants the non-catalytic approach to eliminate NO_x is preferred. This is the direct injection of aqueous urea or ammonia (NH_4OH) into the furnace at 1148-1423°K (875-1150°C). This method is known as the selective non-catalytic reduction (SNCR) as well as Thermal $DeNO_x$ developed in the United States by Exxon [36] and involves the reaction between urea (or ammonia) with the nitrogen oxides formed in the combustion process to yield O_2, CO_2 and H_2O (Eqs. 1 and 2). Due to the difficulty to handle and store ammonia (NH_3), urea ($CO(NH_2)_2$) is applied industrially to reduce NO_x. In the SNCR reaction urea reacts with NO and decomposes it (Eq. 2). The reaction mechanism involves the NH_2 radicals which require a minimum temperature for occurrence of the reaction. Otherwise NH_3 cannot reduce the nitrogen oxides and thus the so-called "NH_3-slip" occurs. The undesirable product "unreacted NH_3" decomposes at high temperatures to NO and water. The reaction(s) need a specific temperature window and sufficient reaction time at this temperature. Ammonia is also corrosive at low temperatures. Practical constraints such as the exhaust temperatures, the reaction time and mixing lead to NO_x conversions that lie between 30 % and 75 %.

$$CO(NH_2)_2 + H_2O \leftrightarrow 2NH_3 + CO_2 \qquad (1)$$
$$2NO + 2NH_3 + \tfrac{1}{2}O_2 \leftrightarrow 2N_2 + 3H_2O \qquad (2)$$

The Selective Catalytic Reduction (SCR)

a) **NH_3-SCR**

The selective catalytic reduction (SCR) is a method which is used for conversion of NO_x from exhaust gases into N_2 and H_2O with the aid of solid catalysts. In this method, the reactions selectively occur between the nitrogen oxides and the gaseous reducing agents, typically an aqueous solution of ammonia (or urea) adsorbed on the catalyst surface. This reaction requires temperatures between 350 and 500°C depending on the catalyst composition and geometry. The principal groups of catalysts that have been investigated for the NH_3-SCR technology are:

(a) metal based oxide catalysts, e.g., those containing vanadium or more recently the sulphated zirconium oxides, the Ni based spinels and the perovskite materials,
(b) supported noble metal catalysts, e.g., Pd/Al_2O_3, Pt/Al_2O_3,
(c) metal ion exchanged zeolites, e.g., Cu-ZSM-5, Fe-ZSM-5.

80 % of NO_x-reduction levels are typically reached with the SCR process. In the presence of air (O_2) urea acts like ammonia similar to the SNCR process. NH_3 and its radicals react with NO and NO_2 as represented by the reactions (3) and (4). In the presence of excess oxygen, the side formation of N_2O as shown in the reaction (5) is favoured [28]. The use of NH_3 as a reducing agent for the selective reduction of NO_x with solid catalysts implies that we require a sensitive control system to avoid the emission of NH_3 into the exhaust, the so-called "NH_3-slip". Unreacted NH_3 causes innumerable environmental and health problems.

$$4NO + 4NH_3 + O_2 \leftrightarrow 4N_2 + 6H_2O \qquad (3)$$
$$2NO_2 + 4NH_3 + O_2 \leftrightarrow 3N_2 + 6H_2O \qquad (4)$$
$$4NO + 4NH_3 + 3O_2 \leftrightarrow 4N_2O + 6H_2O \qquad (5)$$

In the case of the NH_3-SCR technology, the catalyst that found large-scale application is the combination of V_2O_5 and TiO_2 supported on a monolith or a wire screen plate. In general, the NH_3-SCR removes between 60 % and 85 % of NO_x using between 0.6 and 0.9 mol NH_3 for 1 mol of

NO$_x$, however, it may leave between 1 and 5 ppm of the unreacted NH$_3$ (slip). The vanadium based catalyst is promoted with WO$_3$ or MoO$_3$ [28]. Under an oxidizing atmosphere the so-called urea-SCR, commercially known as AdBlue technology, is currently employed for the elimination of NO$_x$ emission in diesel engines. This technology is already in common use to control NO$_x$ emissions from stationary power plants and trucks. Urea-SCR is essentially an ammonia-SCR technology where urea is used to produce ammonia that selectively reduces NO$_x$. The urea aqueous solution is injected upstream of the catalyst, where it undergoes thermolysis to produce ammonia. Under high oxygen content the NO is being oxidized to form NO$_2$ which is effectively reduced by NH$_3$. Fig. 2.8 shows a schematic representation of the catalyst position and the urea dosing system in the BlueTec technology. The Urea-SCR has been shown to be highly effective in on-vehicle testing. However, the successful application of the SCR-technology to vehicles requires a periodic refilling of the urea reservoir. Urea-SCR systems consume urea in proportion to NO$_x$ and can be as high as 5 % of the engine fuel consumption [37]. In addition, key issues that remain to be solved for this technology include; the by- formation of toxic products during the SCR process, the dosing control of the reductant and the durability of the different components of the NO$_x$ control system.

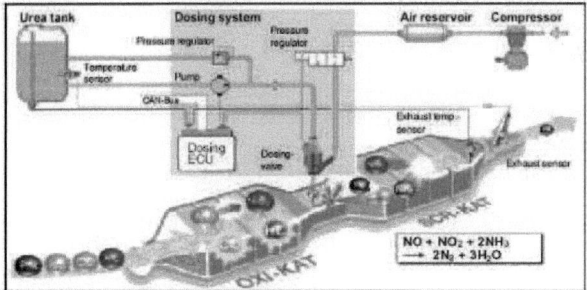

Figure 2.8 Schematic diagram of the Adblue technology for the treatment of emissions in diesel engines (source Bosch).

b) The Selective Catalytic Reduction with hydrocarbons (HC-SCR)

The first technological approach to eliminate NO$_x$ by using hydrocarbons as reducing agent was suggested by Bosch et al [38]. The reduction of NO$_x$ by hydrocarbons is a powerful stimulus for research since unburned hydrocarbons in the exhaust can be used as a reductant during the Selective Catalytic Reduction (SCR). The first catalyst reported to possess HC-SCR activity was based on Cu-ZSM-5 zeolites [39]. Following this, many other cation exchanged zeolites (e.g., Co, Ni, Ni, Cr, Fe, Mn, Ga, In) have been reported to be active in the HC-SCR reaction [40]. Different oxides and platinum group metal catalyst (e.g., Al$_2$O$_3$, TiO$_2$, ZrO$_2$, MgO, and these oxides promoted by, e.g., Co, Ni, Cu, Fe, Sn, Ga, In, Ag compounds) have been also proposed as promising supports [2]. Particularly interesting is the use of CH$_4$ as a reducing agent for the NO$_x$ reduction under excess oxygen (lean-burn conditions). Methane is a highly abundant gas, is also cheap, and could replace ammonia as a reducing agent. The successful use of methane as a reductant in the HC-SCR reaction with a Co-ZSM-5 catalyst was first achieved by Li et al [41].

The use of hydrocarbons (HC) in the SCR technology instead of ammonia or urea seems to be promising. However, the use of hydrocarbons sets up many challenges. For instance, the by-product formation over different catalysts is another problem that remains to be solved. In some cases, almost up to 50 % of the converted NO_x may be reduced to N_2O, which is considered a greenhouse gas [37]. For diesel engines, the engine-out HC is rarely high enough to reduce NO_x and requires active measures such as the injection of hydrocarbons into the exhaust. Increased hydrocarbon concentration may cause a fuel consumption penalty that is compensated with the engine efficiency working under lean-burn conditions. Narula et al have reported an estimated cost increase due to the increase in fuel consumption or equivalent cost of the reductant compared with the NO_x-reduction. [37]. Their calculations show that less than 50 % NO_x-reduction efficiency is achieved at the expense of 2 % fuel consumption increase by using the Pt or Ag supported catalysts.

c) **The H_2-SCR of NO_x**

Since 70 % of the total emissions from a vehicle are produced during "the cold start" phase of the engine operation [9] the future catalyst developments will have to cover this issue. H_2-SCR of NO_x might be a technology that can be applied for NO_x elimination from exhaust engines at low temperatures. Although a relatively large amount of information of the H_2-SCR of NO_x technology exists, the experience has shown that more basic and applied research in this area is needed. The use of hydrogen as a reductant for the NO decomposition in the absence of oxygen over a catalyst with perovskite structure has been rarely studied [42]. The NO-reduction with hydrogen as a reductant under lean conditions (excess oxygen) has been studied mainly over precious metal supported catalysts [4] and ref. therein. For example, platinum supported catalysts are very active for $NO/O_2/H_2$ reactions at low temperatures (below 200°C) but as side effect, large volumes of N_2O are produced. N_2O was the main product at 90°C of the maximum NO-conversion over a 1 wt. % Pt/SiO_2 catalyst. The same behaviour was observed over a 1 wt. % Pt/Al_2O_3 catalyst at 140°C, (the employed feed composition was 200 - 1000 ppm NO + 2000 ppm H_2 and 6 % O_2). Therefore, especially for automotive H_2-SCR NO_x catalysts, an improved N_2-selectivity needs to be targeted.

To date we observe that the most of the progress in the H_2-SCR technology has been made in the development of the precious metal supported catalysts. This is due to the activity of platinum supported catalysts at temperatures below 200°C. During the H_2-SCR reactions no catalytic tests for substituted perovskites have been reported so far. Platinum supported perovskites have been already tested for the H_2-SCR reaction by Costa et al [6]. They observed high N_2-selectivities (~ 90%) over a 0.1 wt. % $Pt/La_{0.7}Sr_{0.3}Ce_{0.1}FeO_3$ catalyst calcined at 400°C. The catalyst maintained a maximum NO_x-conversion after 20 h under steady state conditions. One of the issues of the platinum supported catalysts is the relatively narrow operating window (see Fig. 2.9). High NO_x-conversions are maintained over the 1 wt. % $Pt/La_{0.7}Sr_{0.3}Ce_{0.1}FeO_3$ catalyst at temperatures only between 140°C and 190°C. At higher temperatures the NO_x-conversion strongly decreases.

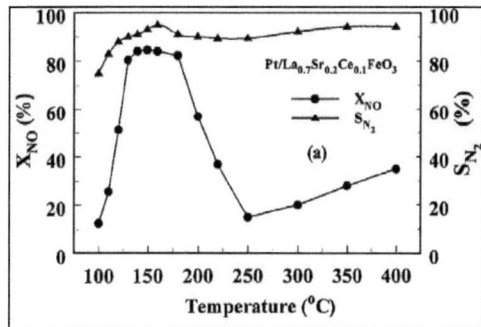

Figure 2.9 NO conversion and N_2-selectivity over 0.15 g of 0.1 wt. % $Pt/La_{0.7}Sr_{0.2}Ce0_{.1}FeO_3$ catalyst. Reaction conditions: 0.25 vol. % NO + 1 vol. % H_2 + 5 vol. % O_2 und 5 vol. % H_2O + He as balance at a GHSV = $80000\ h^{-1}$, reported by Costa et al [6].

d) Lean NO_x trap catalysts

Some lean operating gasoline or diesel engines are equipped with NO_x traps. In a NO_x-trap system the nitrogen oxides are chemically adsorbed on a catalyst, typically BaO. During the lean-burn operation the NO is firstly oxidized in a Pt based catalyst to NO_2 and then reacts with the BaO to form a $BaNO_3$. Due to the presence of excess oxygen, other emissions such as CO, H_2, and unburned HC are oxidized to water and CO_2. The adsorbed NO_x are then reduced to molecular nitrogen under the rich cycle over a suitable catalyst such as those supported noble metals. Fig. 2.10 shows a schematic presentation of the NO_x adsorption-reduction mechanism. Properties of the NO_x-trap, such as the basicity of the storage component, determine the amount of the NO_x adsorbed on the catalyst surface. On the other hand, the basicity of the NO_x-traps affects the HC oxidation capacity. The HC oxidation decreases when a too strong basic NO_x storage element is used [43]. The particle size of platinum directly affects the NO_x adsorption and the reduction properties of the catalyst. In the $Pt/Ba/Al_2O_3$ catalysts with app. 20 wt. % Ba, the smaller the particle size, the higher the NO_x-conversion [43]. For smaller platinum particle sizes, the number of active sites which allows the NO_x-storage would increase. Similarly, the interface between the NO_x-storage phase and the NO_x-reduction active sites (the platinum sites) would increase.

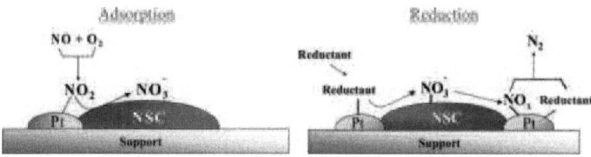

Figure 2.10 Mechanism of the NO_x-storage and reduction with the NO_x-trap technology, M. A. Gómez-García et al [28].

In dynamic systems, the continuous gas treatment requires at least two absorbers that work parallel to alternate between the sorption and the regeneration step. For vehicular applications the trapping or lean-phase is typically 60 - 90 s long, while the regeneration or rich-phase is of the order of 3-5 s [44]. Mechanically the system is complicated and limits the practical application in the exhaust treatment of this process. A second aspect to be solved is the strong deactivation that the

NO_x traps suffer in the presence of small sulphur volumes in the fuel and oil from the engine. Sulphur is oxidized to SO_2 over the platinum sites and then, due to the strong basicity of the barium oxides (BaO) tend to form sulphates. $BaSO_4$ compounds are extremely stable and do not decompose even at temperatures up to 1000°C. The spontaneous decomposition of the $BaSO_4$ would only occur at temperatures higher than 1393°C [45]. For less than 30 % of the BaO involved in the sulphate formation (deactivation), it is possible to regenerate the catalyst under reducing conditions recovering the initial storage capacity even in the presence of remaining sulphates. For a higher poison degree it is more difficult to regenerate the trap and the initial NO_x-storage capacity cannot be recovered, even after a long time reduction treatment [46].

e) Non-thermal plasma (NTP) assisted catalysis for NO_x reduction

The non-thermal plasma is a partially ionized gas state in which the electrons have more energy than the bulk gas. The non-thermal plasma created in an engine exhaust stream combined with a suitable catalyst can be applied to reduce NO_x to N_2, particularly for the activation of the hydrocarbon molecules that are present in exhaust engines, and NO_x-reduction at low temperatures (< 200°C). The plasma assisted catalysis is a two step process consisting of a plasma pre-treatment of the exhaust before flow over a lean NO_x catalyst. The oxidation of the hydrocarbons is partially made in the plasma and subsequent the NO_x-reduction over a catalyst; as a consequence this technology is a type of HC-SCR. In the first step the NO oxidizes to NO_2. In the plasma the hydrocarbons are partially oxidized with the intermediate formation of aldehydes, which are believed to be a key intermediate for the NO_x-reduction reaction. In a second step NO_x are converted into N_2 over the catalyst while the hydrocarbons are consumed [47]. The use of zeolites is a promising alternative to reduce NO_x at temperatures below 200°C with the NTP technology. Generally, the oxide based catalysts show a lower NO_x-reduction compared with the NO_x-conversions on zeolites [48]. Drawback of this technology is the electrical power required to generate the plasma which results in a fuel penalty. A maximum 5 % fuel consumption penalty was set to apply the NTP technology for NO_x-reduction under lean conditions at low temperatures (< 200°C) in a project made by the R & D division of SIEMENS.

2.3 Materials applied for NO_x-reduction under lean conditions

a) The zeolite group of catalyst

The Zeolite group of materials has been extensively studied and reports in the scientific literature proved its potential as NO_x-catalysts under lean conditions HC-SCR [49]. Hydrocarbons such as: methane, propane, propene, butane, etc have been employed as a reducing agent for the NO_x-reduction with zeolites. The use of the hydrocarbons as reductants for the catalytic reduction of NO_x is of practical importance because the unburned hydrocarbons in the exhaust emissions could be applied to reduce NO_x. Handling and transportation issues of NH_3 were solved with the introduction of the urea-SCR technology, but the "NH_3-slip" is still present (see section 2.2). Therefore extensive work has proceeded to search for alternative reducing agents, such as hydrocarbons or hydrogen.

The Cu-ZSM-5 zeolite was the first catalyst reported in the literature to display HC-SCR activity [39]. The discovery that Cu-ZSM-5 zeolite was effective in the NO decomposition and at the same time that this material was more active in the NO reduction with non-methane hydrocarbons promoted a new research field for the NO_x-reduction after-treatment technology. Later studies made clear the thermal and hydrothermal instability of the Cu-ZSM-5 Zeolite catalyst, thus limiting its application in the NO decomposition [50]. Further studies showed that the Co-ZSM-5 Zeolites were effectively applied in the HC-SCR reaction with methane (CH_4) as a reducing agent [41]. Since natural gas, which contains > 90 % methane, is widely used as a fuel source for many combustion processes and electric utilities and is readily available in most parts of the world, its use as a reductant for the NO_x-elimination would be desirable for industrial and automotive applications. Desai et al [51] studied the catalytic properties of the cobalt- and copper- substituted ZSM-5 zeolites during the NO_x SCR reaction under lean conditions employing methane as reductant. The Co-ZSM-5 zeolites showed a relatively high catalytic activity but in the presence of water vapour the NO_x-reduction decreased by a factor of four. The Cu-ZSM 5 is a poor catalyst because CH_4 preferably reacts with O_2 rather than with NO. Water vapour had a dramatic negative effect upon the NO-reduction activity also on gallium loaded zeolites. Upon the addition of 2 vol. % H_2O_{vapour} the NO conversion decreased from 40 % to 13 % at 500°C [49].

Cobalt substituted zeolites reduce NO with CH_4 without exhibiting any strong dependence on larger levels of oxygen up to 21 % [52]. This is of practical importance since the increase in oxygen during combustion process would increase the fuel economy. Table 2.5 reported by Li [52] displays a variety of cobalt substituted zeolites tested for the NO-reduction with CH_4 under excess oxygen. The catalytic performance was strongly dependant on the zeolite type and the metal loading. The direct impregnation of cobalt on oxide supports produced ineffective catalysts. There was a clear synergy between the metal and the zeolite that produces an active catalyst for the NO-reduction.

Armor et al [52] described the effects caused on the NO-reduction performance when using other hydrocarbons such as propene (C_3H_6). Cu- and Co- substituted zeolites posed very similar catalytic activities for the NO reaction under excess oxygen with C_3H_6 as a reductant. In the absence of oxygen the Cu-ZSM-5 catalyst faced a sort of deactivation, probably the coke formation [53, 54]. The NO_x-conversion was improved over a Cu-ZSM-5 zeolite with increased water vapour concentration and decane as a reductant in contrast to propane (C_3H_8) where the catalyst performance decreased [55].

Table 2.5 NO conversion over cobalt exchanged zeolites and cobalt oxide at temperatures 400°-500°C, reported by Li et al [52].

Sample	Metal loading (wt. %)	Conversion to N_2		
		400°C	450°C	500°C
Co-ZSM-5	4.0	23	34	30
CoO/Al_2O_3	11.0	nd	nd	nd
CoO/TiO_2	2.6	nd	nd	nav
$Co/TiO2$	10.0	nd	nd	nd
CoO/Silicalite	1.6	6	5	nav
Co/SiO_2-Al_2O_3	3.0	nd	nd	nd
Co_3O_4	-----	nd	nd	nd

Reaction conditions for all the catalyst tests: GHSV = 30000, 0.1 g catalyst, TFR = 100 ml.min^{-1}, 0.16 vol. % NO, 0.10 vol. % CH_4 and 2.5 vol. % O_2.
nd: conversion not detected
nav: data not available

b) Silver supported group of catalysts

The group of Ag° supported catalysts have shown a very interesting catalytic activity towards the reduction of NO_x with CH_4 and higher hydrocarbons, under relatively high oxygen concentrations. For this type of catalysts the main preparation route is the wet- incipient impregnation of alumina supports with different amounts of silver nitrate solutions [56], after that, the catalysts are usually dried and treated at temperatures between 500° and 650°C. Support properties like the pore-size distribution and the particle size play a very important role in determining the NO_x-reduction activity. There are other chemical routes for the syntheses of these class of catalysts i.e. the precipitation route or the sol-gel method. She et al [57] prepared Ag-alumina materials by the co-gelation technique. The aluminium nitrate and the silver nitrate were used as precursors in these experiments. The two nitrate salts were dissolved in deionised water and precipitated with an aqueous solution of ammonia. The silver content is usually selected to be 2 wt. % in Ag supported catalysts which was reported to correspond to the catalyst composition with the highest activity [58].

The NO_x reduction ability of the silver supported materials has been made over a number of reaction conditions with different hydrocarbons at a range of temperatures from 400°C to 700°C [57]. She et al [57] tested the NO_x-reduction activity of Ag/Al_2O_3 powders with CH_4. The reaction conditions for the CH_4-SCR experiment were 0.25 % NO, 2 % CH_4, and 5 % O_2 in He with a space velocity of 9000 h^{-1}, the catalysts were tested at different temperatures from 300° to 700°C. Over Al_2O_3 the reduction began at 450°C and reached a maximum conversion to N_2 of 70 % at 650°C. With the addition of 0.8 wt. % of Ag to the alumina support, the maximum conversion of NO to N_2 shifted to 600°C and reached to 95 %. An increase in the silver content resulted in a higher conversion of NO and a wider active temperature window were observed, i.e. NO to N_2 conversion reaching to ~100 % at 550°C - 650°C. The authors proposed the reaction pathway for the NO_x reduction with CH_4 as follows: first, methane oxidises by the adsorbed oxygen on silver nanoparticles at 300°C to yield the CO_2 formation. Then, the SCR-active NO_x species are adsorbed on the alumina sites. The production of N_2 starts at ca. 300°C probably due to the interaction between the adsorbed NO_x and the CH_x species. Fig. 2.11 shows the schematic diagram of the mechanistic pathway proposed by She et al [57].

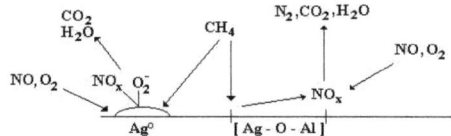

Figure 2.11 Schematic diagram of the reaction mechanism during the selective catalytic reduction of NO with CH_4 over Ag-alumina proposed by She et al [57].

Propene has been also reported as a promising reductant for NO_x-reduction over silver supported catalysts [59]. A 1.2 wt. % Ag/γ-alumina catalyst (BET = 141 $m^2.g^{-1}$) was tested under 0.05 % NO + 0.05 % C_3H_6 + 2.5 % O_2 and He mixtures (total flow = 22.5 $ml.min^{-1}$). The results show that the catalyst was active and selective for the N_2-formation between 400°C and 560°C. The catalyst produced significant amounts of N_2O and NO_2 above 565°C. Very small amounts of NH_3 were also detected; this only before the propene conversion was completed. In addition, the N_2

yield remained significantly lower than that of N_2O and NO_2 formation. Although an acceptable catalytic activity was achieved towards the NO_x-reduction with a variety of hydrocarbons using the Ag-based catalysts, the authors reported that many issues need to be solved such as the deactivation of the catalysts due to the presence of small concentrations of water vapour and SO_2 [56]. This study also reports that the NO_x-conversion to N_2 severely decreased from 62 % to 28 % by the addition of 9 vol. % H_2O and 18 ppm SO_2 into a stream containing 1000 ppm NO, 950 ppm C_3H_6, 10 vol. % O_2, with He balance, at a total flow rate of 500 ml.min^{-1}. The use of alkanes with longer chains has been proposed to increase the water tolerance during the NO_x-reduction reaction. Kenichi et al [60] presented the results concerning SCR experiments with n-octane (n-C_8H_{18}) as a reductant over the 2 wt. % Ag-Al_2O_3 (190 m^2/g) catalysts. One of the proposed reasons for the enhancing effect of water vapour was the inhibition of the unselective oxidation of n-octane, which result in the promotion of the SCR reaction.

c) Platinum group of catalysts

The reduction of nitrogen oxides with methane is also possible by employing Pt supported catalysts. Burch et al [61] prepared Pt/alumina based catalysts for NO-reduction using CH_4. The catalytic activity of the Pt based catalyst (1 wt. % Pt/Al_2O_3 and 1 wt. % Pt/SiO_2) was tested towards the reduction of NO by CH_4 with and without oxygen at temperatures between 200° and 400°C at a total flow rate of 200 cm^3.min^{-1}. A drop in the NO conversion with an increase in the NO:CH_4 ratio was reported. Over the platinum supported catalysts substantial amounts of N_2O were detected at temperatures below 300°C and NH_3 at higher temperatures. Under oxidizing conditions at 270°C, the NO-conversion increases in the presence of small concentration of O_2. During the NO-reduction reaction, the formation of N_2O was observed. The N_2O-formation was strongly dependent on the temperature and the feed composition. At lower temperatures the selectivity to N_2O increased to a maximum as the NO:CH_4 ratio increased [61].

The increase of the N_2O concentration was associated with an increase in the amount of non-dissociative adsorption of NO with the increment in its concentration in the feed [61]. The molecular adsorption of NO seems to be favoured at higher NO pressures. The decline in the N_2O-selectivity with the decreasing CH_4 concentration was proposed to be due to the insufficient adsorption of molecular NO which lowers the rapid coupling of N_{ads} with NO_{ads}. The selectivity to NH_3 formation increased with the increasing CH_4 concentration. As the NO:CH_4 ratio decreased, more adsorbed hydrocarbon species are available to react with the adsorbed oxygen and so the partial combustion of CH_4 would be favoured [61]. More H_{ads} are therefore present under steady state conditions, as so more NH_3 can be formed through the coupling of H_{ads} and N_{ads}. Increase in the oxygen concentration resulted in decline of the NH_3 selectivity. The higher oxygen partial pressure produces statistically a major concentration of O_{ads}, thus reducing the possibility for H_{ads}-formation and hence less NH_3 is produced. Moreover, the excess formation of NO_2 was detected for a gas mixture containing 1000 ppm of O_2 [61].

The platinum supported catalysts showed a moderate activity and a large formation of N_2O with C_3H_6 as a reducing agent during the HC-SCR of NO_x [1]. In comparison the Pd- and Rh- supported catalysts were less active for C_3H_6-SCR of NO_x reactions. However both catalysts showed much lower formation rates of N_2O. Pd/Al_2O_3 showed lower N_2O formation rates than the Rh- and Pd- supported SiO_2 catalysts when applying C_3H_6 as a reducing agent for NO_x-reduction

[1, 62]. These NO_x-conversion values are compared and discussed with the results obtained in the present investigation in the chapter 6.

2.3.1 Perovskites as catalysts for NO_x-reduction

a) Perovskite crystal structure

The Perovskite-type mixed oxides are represented by the general formula $ABO_{3\pm\delta}$, wherein the larger ionic cation A has a dodecahedral co-ordination and the smaller cation B has a six-fold coordination (Fig 2.12). A-cation and oxygen form a cubic closest packing, and B is contained in the octahedral voids in the packing. If the ionic radii are r_A, r_B and r_O, to form a perovskite crystal structure, the tolerance factor $(t)=(r_A-r_O)/(r_B+r_O)2^{1/2}$ must lie within the range $0.8 < t < 1.0$, and $r_A > 0.090$ nm, $r_B < 051$ nm [63]. Both A and B cations can be partially substituted, leading to the general formula $A_{1-x}A_x{'}B_{1-y}B_y{'}O_{3+/-\delta}$. The partial substitution of A' for A of different oxidation states leads to the formation of vacancies of the crystalline bonds and also to stabilization of unusual oxidation states of the cation at the B-site. The B partial substitution by a B' cation can promote changes on both the redox ion couples and the active sites, as well as on the stability of the crystalline structure [64], even at higher temperatures. The physical-chemical properties of perovskites can be tailored, as a large number of the metallic elements can be employed in different amounts and combinations to yield the perovskite structure. The A site of the Perovskite structure can be filled with cations such as; La^{3+}, Sr^{2+}, Ca^{2+}, Ba^{2+}, Nd^{3+}, etc. The B-site can be occupied with smaller cations, i.e. Co, Mn, Ni, Cr, Fe, Al, Cu, etc and the precious metals (i.e. Pd, Pt, Rh). So the multiple combinations of these elements in the same structure may lead to different physical-chemical properties, which make them very interesting materials for catalytic applications. A classical example is the vacancy control in $LaCoO_3$ perovskites by substitution of the lanthanum site with Sr^{2+}, tetravalent Co is produced. Since a tetravalent Co is abnormal in valence, it tends to be reduced by releasing oxygen from the perovskite lattice as shown in the equation 1.

$$La^{3+}_{1-x}Sr^{2+}_xCo^{3+}_{1-x}Co^{4+}_xO_3 \leftrightarrow La^{3+}_{1-x}Sr^{2+}_xCo^{3+}_{1-x+2\delta}Co^{4+}_{x-2\delta}O_3 + \delta/2O_2 \qquad (1)$$

The equation (1) indicates that in $LaCoO_3$, the valence of Co is controlled by the substitution of the lanthanum-site with Sr, leading to an increased reducibility or an oxidizing ability in $La_{1-x}Sr_xCoO_3$ as reported by Tanaka et al [63]. In the case of $LaNiO_3$, as the trivalent La is stable, the nickel ions at the B-site exist not in its usual bivalent state but in the trivalent state. Thus, the perovskite-type oxides are flexible materials which incorporate various metal ions in their stable structures, and can give rise to abnormal valences as well as vacancies by the substitution of the constituent elements.

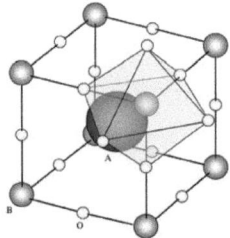

Figure 2.12 Crystal structure of the perovskite type oxide (ABO_3), O = Oxygen [63].

The advantages of the perovskites as catalysts are summarised below [63]:
1. Wide variety of composition and constituent elements keeping essentially the basic structure unchanged.
2. The bulk structure can be characterized well. The surface can be fairly well estimated taking advantage of this well-defined bulk structure.
3. Valences, stoichiometry and vacancies can be varied widely.
4. Huge information on the physical and the solid-state chemical properties has been accumulated.

Basic strategies of designing Perovskite catalysts for the enhancement of their catalytic performance;
1. selection of the B-site element,
2. valence and vacancies controls,
3. synergistic effects mainly of the B-site elements,
4. Enhancement of the surface area by forming fine particles or dispersing on supports, and addition of precious metals.

b) The use of hydrocarbons as reductants for NO elimination over perovskite catalysts

As mentioned in section 2.2 the use of hydrocarbons as reducing agents for NO_x elimination under lean-burn conditions (HC-SCR of NO_x) over supported catalysts has been widely reported in the literature [1, 2, 59]. However only a few work dealing with the HC-SCR of NO_x over catalysts with the perovskite structure exist [65, 66]. Palladium substituted perovskites (the perovskites with palladium in the B-site of the crystal lattice) has been rarely studied for the HC-SCR of NO_x. Zhang et al [3] reported the catalytic behaviour of $LaFe_{0.97}Pd_{0.03}O_3$ catalyst during the C_3H_6-SCR of NO_x reaction. This work reveals that surprisingly higher NO_x-conversions were possible for the cupper substituted perovskite $LaFe_{0.8}Cu_{0.2}O_3$ rather than the palladium substituted perovskite $LaFe_{0.97}Pd_{0.03}O_3$.

There is a need for catalysts which are less selective to the oxidation of the reductant and more selective for the NO-conversion into N_2 under lean-burn conditions (excess oxygen). A second but not less important aspect is the selectivity of the catalyst since lot of N_2O is produced during the reduction of NO_x under excess oxygen [65]. In general, the platinum supported catalysts are very active during $DeNO_x$ reactions but at the same time produce a lot of N_2O as mentioned above. The challenge is to eliminate both NO_x and N_2O under the coexistence of O_2, CO_2, H_2O and SO_2 which are present in the exhausts of automobile engines. The $Pd/LaCoO_3$ based perovskites decompose N_2O in the absence of oxygen, [67, 68]. It is also suggested in the literature that the perovskite catalysts might eliminate N_2O even under excess oxygen [69]. Complete decomposition of N_2O is reported under the coexistence of 10 vol. % O_2 at 627°C over the $La_{0.7}Ba_{0.3}Mn_{0.8}In_{0.2}O_3$ catalyst and a total flow rate (TFR) of 20 ml.min^{-1} (W/F = 3.0 g.s.ml^{-1} ratio). Moreover, it is reported that Ru-supported catalysts i.e. 1 wt. % Ru/Al_2O_3 decompose N_2O completely at temperatures between 450°C and 500°C in the presence of 10 % H_2O, 50 ppm SO_2 and 5 % O_2 at a TFR = 500 ml.min^{-1} and 1 g of catalyst yielding a W/F = 0.12 g.s.ml^{-1} ratio [70]. It is postulated that the oxygen desorption is probably the reaction limiting step after the dissociative adsorption of N_2O in the presence and absence of SO_2 and H_2O.

c) Self healing of perovskites doped with precious metals

The conventional catalysts for automotive applications are composed of finely precious-metal particles dispersed on a solid matrix (i.e. Pd/Al_2O_3, Pt/SiO_2, etc). During the vehicle use, the catalyst is exposed to high temperatures (up to 1000°C) for short time periods that cause the metal particles sintering, the surface area and the catalyst active sites are reduced and thus its catalytic performance as well. An innovative solution that hinders sintering of the precious metal increasing the life of the catalytic converter is by stabilizing the precious metals i.e. Pd, Rh and/or Pt in perovskite mixed oxides. The stabilization of the precious metal(s) in perovskitic crystal structures provides a better thermal resistance of the catalyst [12]. Moreover precious metals in perovskite structures pose self-healing properties, as an example palladium diffuses out of the crystal lattice under rich-conditions and is reincorporated in the frame lattice under lean-conditions in lanthanum based perovskites as shown in Fig. 2.13 [12]. Interesting is that the reversible redox behaviour of palladium in perovskite crystals is coupled with the functioning of automotive gasoline engines that operate close to the stoichiometric point ($\lambda=1$). To maintain the stoichiometric conditions during combustion, oxygen sensors provide the feedback information needed to efficiently eliminate CO and NO. Fluctuations in the exhaust gases between rich and lean conditions occur due to adjustment of the air-to-fuel ratio. The palladium integrated perovskite uses the redox fluctuations in the exhaust gases for a self-healing function of the catalyst providing a better performance in the long term use of the catalyst [12].

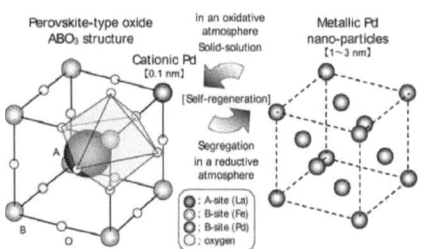

Figure 2.13 Principle of the self-healing of the palladium integrated perovskites [71].

A wide variety of perovskites prepared by the alkoxide method stabilizes precious metals [12]. For instance palladium forms a solid solution with the $LaFeO_3$ and $LaFe_{(1-x)}Co_xO_3$ based perovskites, and the metals such as rhodium forms a solid solution with the $LaFeO_3$, $LaAlO_3$, $CaTiO_3$, $CaZrO_3$ oxides, while platinum forms a solid solution with $LaFeO_3$, $CaTiO_3$, $CaZrO_3$, $SrTiO_3$ and $BaTiO_3$ perovskites [11]. So far no information regarding the possibility of a solid solution between $BaTiO_3$ and palladium was reported. An interesting observation is that the $CaTiO_3$ perovskite forms a solid solution with rhodium and platinum, and no single phase was formed with palladium [11].

Another important result is that the diffusion degree of the precious metals strongly depends on the host perovskite. A large difference in the proportion of the precious metal present in and outside the perovskite lattice defines the self-regeneration ratio. Different perovskites under redox fluctuations may yield large differences in the self-regeneration ratio. Based on the coordination number of the first nearest neighbour with oxygen (Pd-O, Pt-O bonds), it is possible to estimate the diffusion degree of the noble metal in and out of the perovskite. For instance for the $LaFe_{0.95}Pd_{0.05}O_3$ perovskite a complete diffusion of palladium and for the $CaTi_{0.95}Pt_{0.05}O_3$ perovskite an incomplete diffusion have been determined [72].

The palladium substituted perovskite $LaFe_{0.57}Co_{0.38}Pd_{0.05}O_3$ and the Co-free parent $LaFe_{0.95}Pd_{0.05}O_3$ are the most studied materials from the group of perovskites mentioned above [11, 12, 73, 74]. Under reductive atmosphere palladium segregates out of the crystal lattice from the perovskite $LaFe_{0.57}Co_{0.38}Pd_{0.05}O_3$ to form a metallic alloy with cobalt, a disordered face-centred cubic (fcc) structure containing Pd and Co [73]. Iron pose two important functions, stabilizes the perovskite structure and suppresses the growth of the palladium particles. The iron-free perovskite $LaCoO_3$ decomposes into $Co°$ and La_2O_3 at temperatures above 300°C under strong reduction conditions [75]. The decomposition mechanism of this perovskite is still not well understood. Under redox cycle 30 % of the elements at the B-site of the perovskite $LaFe_{0.57}Co_{0.38}Pd_{0.05}O_3$ are reduced into its metallic form. The elements at the B-site of the perovskite were reduced in the order: Pd>Co>Fe. Palladium in the $LaFe_{0.95}Pd_{0.05}O_3$ perovskite started to segregate at 100°C, about 30 % of palladium was converted into $Pd°$ at 200°C and approx. 90 % at 400°C [74], while only a small quantity of Fe was transformed into its metallic state, only a small fraction of a Pd-Fe alloy was observed. The amount of palladium had no effect on the final perovskitic structure. The complete solid solution of the $LaFe_{(1-x)}Pd_xO_3$ (x = 0.1) perovskite was implied in the literature [76]. On re-oxidation all the reduced elements of the perovskite were reincorporated in the frame of the perovskite lattice. In addition to the self-healing properties of the perovskite $LaFe_{0.95}Pd_{0.05}O_3$, this catalyst showed an oxygen storage capacity (OSC) that was comparable to the OSC of the cerium based oxides [74].

The palladium doped perovskites $LaFe_{0.57}Co_{0.38}Pd_{0.05}O_3$ and $LaFe_{0.95}Pd_{0.05}O_3$ are single crystal phases reported to be thermally stable up to 900°C under oxidizing conditions, even after ageing for 100 h under real exhaust gases from a gasoline engine [73]. A very small amount of the $La(OH)_3$ phase according to the XRD intensity signals in the pattern of $LaFe_{0.57}Co_{0.38}Pd_{0.05}O_3$ after the reduction in 10 vol. % H_2 / 90 % N_2 at 800°C/1h was reported [73], no other phase(s) were found after the calcination under oxidation conditions. The phase stability of the palladium substituted in perovskitic structures also depend on the host elements as mentioned before [77].

2.4 The EB-PVD coating method

The Electron Beam Physical Vapor Deposition (EB-PVD) is a form of physical vapour deposition in which a target anode is bombarded with an electron beam giving off by a charged tungsten filament under high vacuum. The electron beam following a scanning pattern upon the surface of the water-cooled ceramic target causes atoms to transform into the gas phase. These atoms then condensate on the surface of previously pre-heated substrates fastened to a sample holder positioned inside the vapour cloud. Fig. 2.14 displays the schematic diagram of the EB-PVD process. This coating method provides high deposition rates (few micrometers per minute) offering structural and morphological control of the film. Moreover, it is possible to produce coatings on turbine blades by leaving the cooling holes free, providing less machining requirement. For these reasons this process has a potential for industrial applications for wear resistant and thermal barrier coatings in aerospace industries, hard coatings for cutting and tool industries, and electronic and optical devices. The field of industrial applications of the EB-PVD coating method can be extended to the development of ceramic films for automotive catalysts.

The EB-PVD coating process begins with the nucleation of the vapour phase on the preheated substrates. Small grains with no preferred crystal orientation are formed initially at the bottom of the coating substrates [78]. The growth process of individual columns occurs in a preferred crystallographic orientation governed by the developed rooftop configuration of the column tips. The enlargement of the columns diameter and elimination of non-favourable oriented columns occur due to a process influenced by the "sunrise-sunset" line of sight shadowing effect of the neighbouring columns during the sample rotation. The rotation of the substrates produces the mentioned "sunrise-sunset" vapour incidence pattern that determines the deposition rate over a differential area as a function of time. Thus, the interplay between neighbouring column tips and the vapour incidence angle of the incoming particles enhances the dominance of the shadowing effect over the surface diffusion. Melting point of the coating material and the temperature of the substrate affect the columns growth during the coating process. B. A. Movchan and A. V. Demchishin proposed the physical model that describes the microstructure of the EB-PVD coatings as a function of substrate temperature (T_s) and the melting temperature of the deposited material (T_m) [79]. This model presents three zones with different coating microstructures, in zone 1 the films are characterized by a low surface diffusion of the deposited atoms on the growing surface. The resulting grains are cone-shaped with rounded tips, enclosing a highly porous (fibrous) cross section. In zone 2, the activation of the surface diffusion defines the appearance of parallel columnar grains with faceted tips, containing lower porosity volumes. Zone 3 is characterized by the additional activation of volume diffusion. The resulting microstructure is composed by re-crystallized equiaxed grains with a flat top surface (see Fig. 2.15). The integration of catalysts to the EB-PVD PYSZ coatings has not been made so far.

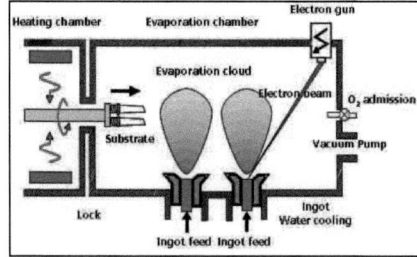

Figure 2.14 Schematics of the main components of the Two-Source Jumping Beam Electron method.

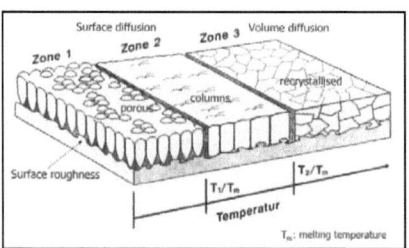

Figure 2.15 Structure zone models showing the effect of the substrate temperature and roughness on the microstructure of the EB-PVD coatings proposed by B. A. Movchan and A. V. Demchishin [79].

2.5 Open questions and material development required for NO_x-reduction under lean-burn conditions applying perovskites

The issues regarding the development of perovskites and in its application as NO_x catalysts are listed here:
  a) Safer synthesis routes,
  b) The quality of the catalyst (reproducibility of the synthesis route(s)),
  c) The HC-SCR and H_2-SCR capability of the perovskitic structures,
  d) The N_2-Selectivity of the catalysts during the HC-SCR and H_2-SCR of NO_x,
  e) The surface functionalizing of the perovskite catalysts for NO_x-reduction under rich and lean environments,

Properties such as the phase stability, surface area, and the catalytic properties (i.e. NO_x-conversion, N_2-selectivity, the resistance against poisoning, etc) strongly depend upon the synthesis route and the factors implied during this process (i.e. pH, water concentration, nature of the reactants, temperature, etc). Many synthesis methods to prepare perovskites involve the use of nitrates because they can be easily dissolve in aqueous based solutions providing a well homogenized mixture even at atomic level [80]. The solvent elimination and the calcination of these nitrate(s) solutions produce NO_x gases as a by-product. The use of nitrates as precursors may lead in some cases to the formation of explosive mixtures in particular when cobalt and iron ions are present. As a consequence environmentally friendlier synthesis routes for perovskite catalysts would be needed.

The self healing capacity under cyclic redox environments, makes the perovskite structures very useful as catalysts [12]. The perovskite with the composition $LaFe_{0.57}Co_{0.38}Pd_{0.05}O_3$ maintains its crystal structure with high palladium dispersion even after a thermal aging at 900°C for 100 h under redox environment [63]. Iron in the B-site of the perovskite confers structural stability to this compound even under redox conditions [63]; as the perovskite without iron ($LaCoO_3$) decomposes into La_2O_3 and $Co°$ [75]. The mechanism of this phase transformation(s) is still not well understood. On the other hand the La_2O_3 + PdO mixed oxides are believed to be metastable. Palladium oxide in the mixture La_2O_3 + PdO may decompose into the metallic state at 800°C in air [81]. There exist only a few studies which report the crystal structure stability of perovskites under redox atmospheres. The $La_{0.9}Ce_{0.1}Co_{(1-x)}Fe_xO_3$ (x = 0, 0.2, 0.4, 0.6, 0.8, and 1.0) perovskites prepared via the alkoxide technique split into the $LaFeO_3$, CeO_2 and La_2O_3 oxides after the calcination at 700°C/2h, $LaFeO_3$, CeO_2, La_2O_3, and Co_3O_4 were observed after a treatment at 900°C/4h [82]. A catalyst with the composition $Pd/La_{0.9}Ce_{0.1}Co_{0.4}Fe_{0.6}O_3$ showed the best balance between the catalytic activities for CO/NO reactions and the durability after an aging under cyclic redox environments. The crystal phase(s) present in the catalyst have a direct effect in the NO_x-reduction capability. As the literature overview provided in the chapter 2.3, the supported or the substituted precious metals on perovskite structures offer a wide possibility for new catalyst developments for NO_x-reduction applications.

In general, the TWC which is applied to treat the gas emissions from gasoline engines cannot reduce NO_x under lean conditions. The SCR technology is a promising approach to eliminate NO_x under lean atmospheres at low (< 200°C) and high temperatures (> 200°C). For automotive exhausts of a low temperature the H_2-SCR technology would be an alternative to reduce NO_x in lean exhausts ("Cold start"). To the best of my knowledge the palladium substituted perovskites have not been tested yet for $NO/O_2/H_2$ reactions. For lean exhausts with higher temperatures the HC-SCR method may be an alternative to eliminate NO_x. Only the study of Zhang et al [3] reported the catalytic properties for C_3H_6-SCR of NO_x with a catalyst with the composition $LaFe_{0.97}Pd_{0.03}O_3$ at 500°C. An increasing potential for perovskite structures can be envisaged as automotive catalysts for NO_x-reduction under lean-burn conditions with the use of hydrogen, HC or both as reductant(s). Moreover, precious metals substituted in the perovskites have shown a better thermal resistance compared to the typical supported catalysts, this makes perovskites very attractive as host structures for precious metals. Catalysts with better durability in the long term application would inherently cause a less precious metal consumption and would solve the supply-demand issues of the automotive industry.

As indicated in the chapter 2.3, the selectivity of the catalyst is extremely important for practical applications. The platinum supported catalysts are very active for NO_x-reduction under lean conditions applying C_3H_6 [1] or even with hydrogen [4] as reducing agent(s). However, the platinum supported catalysts produce large amounts of N_2O. Standard limits for N_2O are still not considered by the legislation in the European Union, but it can be foreseen that N_2O will also be regulated in the near future. The use of cerium containing perovskites seems promising for further improvements in the NO_x–reduction performance and N_2-selectivity [5, 83, 84].

Novel concepts composed of one or more units for the NO_x elimination under lean and rich conditions have been proposed in the past. One concept is the TWC catalyst combined with the NO_x-storage, and the oxidation catalysts [22]. More than one catalytic unit is necessary in order to efficiently eliminate NO_x, UHC, and CO_x during the driving cycle of automobiles. The development of catalysts with a NO_x-reduction capability and NO_x-storage properties named as bi-functional catalysts (or even multi-functional catalysts) may provide weight savings and less CO_2 emissions. Perovskites offer the possibility to combine many and different elements preserving the same crystal structure. A practical example is the compound $BaTi_{(1-x)}B'_xO_3$ (B' = Pd, Rh, and/or Pt) which stores NO_x under lean-burn conditions as proved in our study, and under rich conditions the stored NO_x may be eliminated as in the TWC [28]. In addition, the system may function as a SCR catalyst depending on the precious metal substituted at the B'-site of the host perovskite. The adaptive surface properties of a highly active perovskite catalyst would selectively reduce NO_x under lean and rich conditions with additional CO_2 savings due to a weight reduction. But before the development of a catalytic converter, basic research is needed regarding the use of perovskites as catalysts for NO_x reduction under rich and lean-burn conditions paying attention to the selectivity of the catalyst.

Chapter 3

3. Motivation and objectives of the present investigation

Motivation

There is an urgent need for a further reduction of the NO_x-emission by 2010 (i.e. < 0.2 g/km for diesel passenger cars). These limitations are ordered by the stringency of the current international (EPA's Tier 2, Super Ultra Low Emission Vehicles - SULEV) regulations and the imminent European legislations (EURO 5 and 6). The hydrogen powered fuel-cell vehicles are still someway from a commercial production. For the gasoline and diesel powered vehicles, the NO_x-emission occurs mainly by the oxidation of nitrogen during the combustion. Environmental pollution due to the NO_x-emission is a serious problem for environment and human-beings, since it causes acid-rain and increases the ground ozone level. Other pollutants, for instance SO_2-emission, can be significantly reduced, if natural gas is used as a combustion fuel. The NO_x-emissions, however, are not avoidable solely by employing alternative fuels.

The Three Way Catalyst (TWC) has been one of the best suitable technologies for reduction of emissions in the stoichiometric burn engines. Lean-operated engines (i.e. Air/Fuel >14.6) such as "Diesel" or "Gasoline Direct Injection" engines have much higher fuel efficiency. However, the use of TWCs is limited at the exhausts of these engines, since it yields in lean atmospheres only a marginal NO_x reduction. That is why the Selective Catalytic Reduction (SCR) is the preferred technological approach to eliminate NO_x in the presence of high oxygen concentrations. The SCR method can be a highly selective method (to N_2), particularly with the right dosage of reductant e.g. NH_3, HC, H_2. The NH_3-SCR method still has some problems that must be addressed in the near future as explained in chapter 2.2 section (a). Alternatively, hydrocarbons (HC-SCR) and/or the hydrogen (H_2-SCR) are employed to reduce NO_x under lean conditions, see the chapter 2.2 section (b) and (c). The N_2O-formation is still one of the main drawbacks of the SCR technology when employing hydrocarbons or hydrogen as a reductant(s). It can be foreseen that new catalysts with a better NO_x-reduction capacity and improved N_2-selectivity will be needed. In the present work, the synthesis and characterization of catalytic coatings for the selective reduction of NO_x at low and high temperatures are carried out.

Objectives

The main aim of this study is to develop catalytic coatings to eliminate NO_x under lean conditions. The coating development include the synthesis of the catalyst(s) (the active phase) and the coating itself. It is considered the development of an adaptive perovskite based-catalyst that offers higher efficiency in the NO_x-conversion under lean- and rich- conditions. It is planed to stabilize palladium ions in perovskite crystals in order to improve the NO_x-conversions and N_2-selectivities of the catalyst(s). The precious metal based catalyst(s) would be able to function as a Three Way Catalyst (TWC) in order to reduce NO_x under rich-conditions as previously mentioned in the section 2.3.1. At the same time the perovskite(s) would work as a SCR catalyst to eliminate NO_x in the presence of excess oxygen, thus providing a catalyst with adaptive surface properties to the reaction conditions. The development of a new catalyst is envisaged with an improved N_2-selectivity and high NO_x-reduction levels under lean-burn conditions at low (cold-start) and high operation temperatures.

The present investigation contemplates the state-of-the-art of the scientific literature in order to provide an insight into the fundamental mechanisms related to the substitution of doping elements in the perovskitic structures for high levels of selective NO_x-conversion in oxygen-rich environments.

As the particular objectives of the present study; the stability of the perovskite crystal structure and the changes in the catalytic activity caused by the substitution of different elements at the A- and B- sites are studied (ABO_3, A = La, Ba, Ce and B = Fe, Co, Ti). The effect of the doping elements such as ceria, to the NO_x-reduction and N_2-selectivity of the catalysts are investigated. The characterization techniques such as SEM, XRD, TEM and XPS are used to study the substitution of palladium ions in the perovskite lattice, and to demonstrate the reversible diffusion of these substituted metal ions in and out of the La- and Ba- based perovskites under redox conditions.

The present investigation includes the following particular objectives; the use of a modified citrate route to prepare the lanthanum based perovskites integrated with palladium (the catalyst development). The crystal structures and the phase composition modifications with different doping elements such as (Co, Fe, etc) and with increasing calcination temperatures of the perovskite powdered catalysts are studied. A particular attention is paid to the possible self-healing properties under redox conditions of this kind of perovskites. Only selected compositions are intensively analyzed via TEM and XPS in order to prove the Pd-substitution and the reversible diffusion of the substituted palladium ions in and out of the perovskite lattice(s).

The characterization of the perovskite based-catalysts includes the NO_x-reduction measurement under model gas mixtures containing NO_x, high oxygen concentrations and hydrogen as reductant. The quantification of the effects of H_2O_{vapour} + CO_2 and CO to the NO_x–reduction performance and N_2-selectivity of the synthesized perovskites during the H_2-SCR of NO_x reactions is considered. Alternatively the catalytic activity determination for C_3H_6-SCR of NO_x reactions of selected catalysts compositions (La- and Ba- based perovskites) including the effects of increased propene and oxygen concentrations to the NO_x-performance of the catalysts is contemplated.

The points analyzed during the development of the catalytic coating are listed below;
- Development and microstructural characterization via XRD and SEM,
- Integration of the perovskite based catalyst(s) to the EB-PVD PYSZ coatings applying the incipient method,
- Analysis and microstructural evaluation of the EB-PVD PYSZ coating before and after impregnation with the perovskite catalyst,
- Determination of the cyclic thermal ageing effect to the microstructure and the phase composition of the synthesized catalytic coatings,
- Development and microstructural characterization of a perovskite catalyst coated on cordierite substrates (the catalytic converter development),
- And the NO_x-reduction performance determination under lean conditions (C_3H_6-SCR of NO_x) in the presence and absence of water vapour of the perovskite integrated catalytic converter.

Chapter 4

4. Materials and methods

The next section describes the synthesis route employed to prepare the catalysts studied in this work. Basically two different synthesis methods were applied; a modified citrate route to synthesize $LaFe_xCo_{(1-x-y)}Pd_yO_3$ (x = 0.4, 0.475 and 0.65) and the co-precipitation method to synthesize the $BaTi_{(1-x)}Pd_xO_3$ based perovskites. The B-site of each perovskite (Co in the case of $LaFe_xCo_{(1-x-y)}Pd_yO_3$ and Ti in the case of $BaTi_{(1-x)}Pd_xO_3$ was substituted with 5 mol % of palladium. The La-site (A-site) in the $LaFe_xCo_{(1-x-y)}Pd_yO_3$ perovskite was also replaced by small amounts of ceria. The Co-free $LaFe_{(1-y)}Pd_yO_3$ catalysts were also synthesized via the citrate route. The perovskites without palladium were prepared employing the same synthesis method in each case. Similarly, palladium supported samples were synthesized via the direct impregnation sample with Pd-nitrate solution (see Table 4.1). For simplicity the sample coding is used in the following chapters. This section provides a detailed description of the synthesis methods employed in this work.

Table 4.1 Catalyst compositions and coding synthesized and studied in this study [a].

Citrate route	Sample Code	State
$LaFe_{0.475}Co_{0.475}Pd_{0.05}O_3$	$LaFeCo_{(0.475)}$-Pd	
$LaFe_{0.475}Co_{0.475}Rh_{0.05}O_3$	$LaFeCo_{(0.475)}$-Rh	
$LaFe_{0.55}Co_{0.4}Pd_{0.05}O_3$	$LaFeCo_{(0.4)}$-Pd	Precious metal
$La_{0.95}Ce_{0.05}Fe_{0.65}Co_{0.3}Pd_{0.05}O_3$	$LaCe_{(0.05)}FeCo_{(0.3)}$-Pd	integrated
$La_{0.95}Ce_{0.05}Fe_{0.95}Pd_{0.05}O_3$	$LaCe_{(0.05)}$Fe-Pd	catalysts
$La_{0.6}Ce_{0.4}Fe_{0.95}Pd_{0.05}O_3$	$LaCe_{(0.4)}$Fe-Pd	
$LaFe_{0.65}Co_{0.3}Pd_{0.05}O_3$ *	$LaFeCo_{(0.3)}$-Pd	
Pd-$LaFe_{0.65}Co_{0.35}O_3$ *	Pd- $LaFeCo_{(0.3)}$	Precious metal (impregnated)
$LaFe_{0.65}Co_{0.35}O_3$ *	$LaFeCo_{(0.35)}$	No precious metal
Co-precipitation method		
$BaTi_{0.95}Pd_{0.05}O_3$ *	BaTi-Pd	Precious metal integrated catalysts
Pd-$BaTiO_3$ *	Pd-BaTi	Precious metal (impregnated)
$BaTiO_3$ *	BaTi	No precious metal

* Catalysts compositions subjected to main investigations
[a] This table also pops up in appendix 4

4.1 Synthesis of the catalysts.

The present chapter provides a detailed description of the synthesis method(s) employed for the synthesis of the catalysts studied in the frame of this research study. The citrate route was selected to synthesize the substituted mixed oxide(s) $LaFe_xCo_{(1-x-y)}Pd_yO_3$. This method offers the possibility to mix oxides at a wide range of compositions; in addition the stoichiometry of the mixed oxides can be controlled at a very good level. The co-precipitation method was applied to synthesize the group of catalyst(s) with the composition $BaTi_{(1-x)}Pd_xO_3$.

4.1.1 The citrate route

The palladium integrated perovskite $LaFe_xCo_{(1-x-y)}Pd_yO_3$ was prepared employing the citrate route [85]. First, an aqueous solution of Lanthanum acetate (99.9 %) was prepared making the A-site. Separately, Co acetylacetonate (98 %) and Fe nitrate (99 %) were dissolved with distilled water (B-site of the Perovskite structure). Sol formation is provided by adding concentrated nitric acid to an aqueous solution of Co- acetylacetonate. Palladium di-nitrate $(Pd(NO_3)_2)$ was first dissolved in a 30 vol. % HNO_3 aqueous solution and added to an iron nitrate $(Fe(NO_3)_3)$ solution. A stoichiometric amount of citric acid with respect to the B-site of the perovskite was added to the mixture. Citric acid forms a complex compound with the metal cations i.e. Co- and Fe- facilitating the elimination of nitrates as nitrogen oxides. The nitrate precursors are generally used in this kind of synthesis due to their good solubility in water. However problems arise during the solvent elimination and further solution concentration during the drying and the calcination stages of the precursors. The resulting solution was mixed with a La acetate $(LaCOO(CH_3)_3)$ aqueous solution under thoroughly stirring. At this point ethylene glycol was added at a mass ratio of 40:60 with respect to the citric acid. Ethylene glycol or similar organic substances work as a fuel for oxidation reactions that take place during the concentration, the drying and the calcination stages of the solution. Without any organic additives for the perovskite synthesis extremely exothermic reactions take place and the risk of explosion or fire hazard increases notably. This is likely when the metal cations such as Fe or Co are present, which catalyze the oxidation of the organic matter. In our synthesis the solvent was removed with a rota-vapour at 80°C under vacuum. The temperature was maintained constant (80°C) up to almost the complete elimination of the solvent and resin formation. Partial combustion and the elimination of the hydrocarbons was made over a heated plate. A spongy solid was formed and then this solid was slightly grounded with an agate mortar, kept in oven at 80°C for 24 h and finally calcined for 3 h at various temperatures, ranging from 500°C to 900°C.

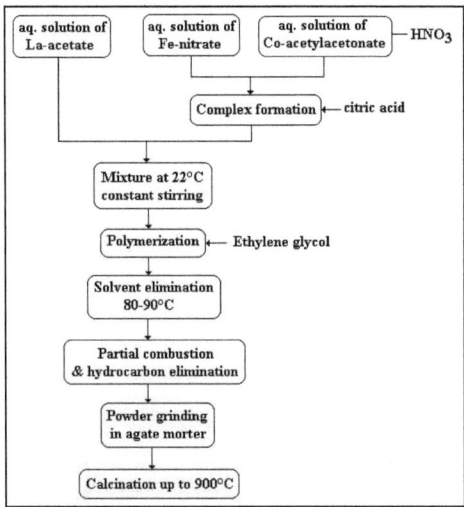

Figure 4.1 Flow diagram of the $LaFeCo_{(x)}$-Pd synthesis route by the citrate route [85].

For comparison purposes the perovskite $LaFeCo_{(0.35)}$ without palladium was prepared using the same synthesis method. Additionally, synthesis of the palladium supported sample was made by direct impregnation of the perovskite $LaFeCo_{(0.35)}$ designated as $Pd-LaFeCo_{(0.35)}$. For the synthesis of the precious metal impregnated catalyst to yield $Pd-LaFeCo_{(0.35)}$, the Pd-free perovskite $LaFeCo_{(0.35)}$ was first treated at 900°C/3h to obtain a well defined perovskite structure, the so obtained powder was then impregnated with a palladium nitrate solution to advice the same Pd content as in the substituted sample. Finally, the catalyst was dried at 70°C overnight and calcined at 700°C/3h for further analysis.

4.1.2 The co-precipitation method

The BaTi-Pd based catalyst was synthesized by a co-precipitation method. A new synthesis method involving the use of barium metallic pieces and titanium isopropoxide was applied. To avoid any extreme exothermic reaction(s), barium metallic basis pieces must not come in contact with aqueous solutions or water at all. On the other side Ti-isopropoxide $Ti[OCH(CH_3)_2]_4$ immediately polymerises in contact with water. Therefore, a solution of the Ti-isopropoxide in 2-propanol was applied which provides the opportunity to obtain an homogeneous mixtures with the barium solution. A homogenous solution is important to avoid any segregation of second phase(s) after the drying and the calcination process. The flow chart of the synthesis route is shown in Fig. 4.2. The barium based solution was prepared by dissolving the barium metallic pieces (metal basis, 99.5 %) in 2-propanol under a constant stirring overnight. Palladium nitrate dehydrate (99.9 %) was added to this solution. Simultaneously, the Ti-based sol was prepared by dissolving a titanium isopropoxide (99.999 %) in 2-propanol. The co-precipitation was achieved by rapid addition of the titanium sol to the Ba-sol previously mixed with the Pd-nitrate solution. The obtained gel was then thoroughly mixed until a homogeneous suspension was formed. After that, the solvent was eliminated at 80°C using a rota-vapour. Finally, the obtained xerogel was heat-treated at different temperatures up to 900°C with a holding time of 3 hours at each temperature.

The synthesis of the Pd-BaTi supported catalysts was achieved by the direct impregnation method. For this purpose, the BaTi perovskite was first treated at 900°C to obtain a well-defined perovskite structure, and then the perovskite powder was impregnated by the direct incipient wetness method. First a palladium nitrate solution was prepared and then the powder BaTi-900°C was added to the solution and the suspension was thoroughly mixed in an ultrasonic bath for 15min. Finally, the resulting suspension was then dried at 70°C overnight and calcined at 700°C/3h (see table 4.1). For comparison, the Pd-free BaTi perovskite was prepared by employing the same synthesis method.

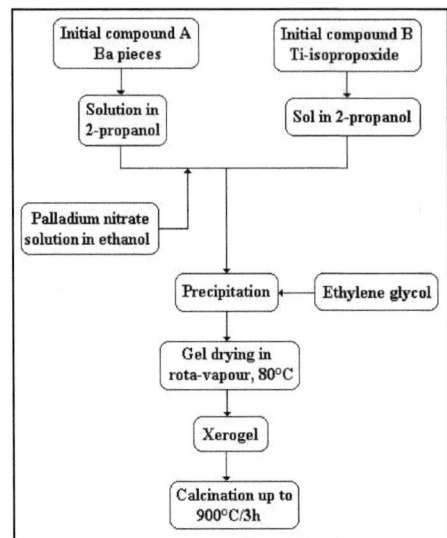

Figure 4.2 Scheme of the perovskite BaTi-Pd synthesis route by the co-precipitation method.

4.1.3 Manufacture of the integrated perovskite coating system.

Standard Al_2O_3 plates for deposition of the EB-PVD PYSZ coatings were employed as model substrates allowing high temperatures resistance during the coating (900° - 1000°C). The alumina plates were firstly cleaned in an ultrasonic bath using distilled water and finally dried with hot air. The substrates were then fixed on a six flat edge rotating horizontal holder aligned perpendicular to the ingot axis. The Al_2O_3 plates (dimensions; 4x10 cm) were coated in a 150 kW (40 kV) pilot plant supplied by "Ardenne Anlagetechnik GmbH" (Fig. 4.3). Al_2O_3 plates coated with EB-PVD partially stabilized zirconium oxide (PYSZ) having a 7 mol % Y_2O_3 as stabilizing oxide were used for the synthesis of the integrated perovskite-EB-PVD coating system.

Alumina plates were homogeneously pre-heated between 920°C - 1000°C in order to eliminate adsorbed gas species (i.e. adsorbed water) from the substrate surface. The substrate temperature was constantly monitored by a thermocouple which was in direct contact with the substrates. The substrate temperature determines the homologous temperature T_s/T_m (K) where T_s = substrate temperature and T_m = melting point temperature of the ceramic to be deposited. In the practice, in order to achieve a well defined columnar microstructure with a smooth surface, T_s is selected to be within the range of $0.3\ T_m < T_s < (0.45 - 0.5)\ T_m$. The oxygen partial pressure in the coating chamber is controlled via a mass flow controller which was adjusted to required vacuum (1×10^{-4} mbar) to obtain the desired stoichiometry of the PYSZ coating. The deposition of the PYSZ coatings was achieved by employing a single evaporation source which had the composition of 7 wt. % Y_2O_3-ZrO_2, the ingot diameter was 63.5 mm. The ceramic ingot was fed into the evaporation/deposition chamber through a water cooled copper crucible. The coating conditions used for the sample preparation in this work provide a deposition rate of 5 - 6 $\mu m.min^{-1}$. Final thickness of the PYSZ coatings was 190 µm (± 10 µm).

Figure 4.3 EB-PVD 150kW (40kV) from Ardenne Anlagetechnik GmbH pilot plant

The dip coating method was employed for the integration of the La- based perovskite with the PYSZ coating. Three different approaches were employed for the impregnation of PYSZ: (The direct impregnation, a particulate coating, a particulate coating plus sol-coating method). The direct impregnation of the EB-PVD PYSZ coating was achieved by dip coating into a $LaFeCo_{(0.3)}$-Pd sol (sol-coating method), followed by a drying and calcination processes. The schematic diagram of the dip coating process is presented in Fig. 4.4. The $LaFeCo_{(0.3)}$-Pd sol was prepared as described in section 4.1.1. The coated EB-PVD PYSZ substrates were dried at room temperature for a few minutes and then calcined at temperatures up to 900°C for 3 h.

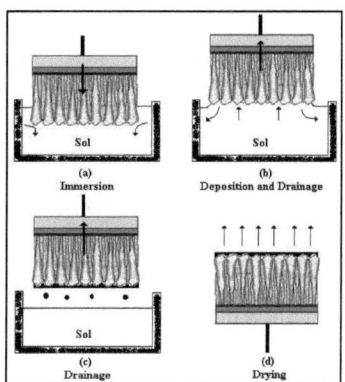

Fig. 4.4 Schematic diagram of the dip-coating process

Similarly, in a second approach (the particulate coating method); the EB-PVD PYSZ samples were coated via the dip-coating process as shown in Fig. 4.4. For coating, the particles of $LaFeCo_{(0.3)}$-Pd were previously synthesized via the citrate route and calcined at 500°C for 3 h and milled in order to obtain a powder with a bimodal particle size distribution (0.1-10µm). These perovskite powder particles were then used to prepare an aqueous suspension. With this approach the perovskite mass on the final coatings was increased. However adhesion between the substrate and the powder particles was not good.

In order to increase the mass of the catalyst on the PYSZ coating surface and to improve the adhesion particle-particle and particles-substrate, a third method is applied which combines these two first impregnation methods (the particulate plus the sol-coating method). A suspension was prepared with 10 % of solid particles LaFeCo$_{(0.3)}$-Pd-500°C/3h in the LaFeCo$_{(0.3)}$-Pd sol. The sol of LaFeCo$_{(0.3)}$-Pd was prepared via the processing route described in section 4.1.1. The solvent from the sol of LaFeCo$_{(0.3)}$-Pd was partially evaporated on a heated plate and the powder LaFeCo$_{(0.3)}$-Pd-500°C/3h was added to the mixture under a vigorous stirring. The mixture was then homogenized for a few minutes and the final coating of the EB-PVD PYSZ substrates was made via the dip-coating process (see Fig 4.4). The EB-PVD PYSZ substrates coated with the LaFeCo$_{(0.3)}$-Pd suspension/sol were dried at room temperature and calcined at temperatures up to 900°C for 3 h. Fig. 4.5 shows (from left to right) the surface appearance of the as prepared EB-PVD PYSZ coating compared with the coatings LaFeCo$_{(0.3)}$-Pd–PYSZ prepared via the combined dip-coating process employing the sol LaFeCo$_{(0.3)}$-Pd + the solid powder LaFeCo$_{(0.3)}$-Pd.

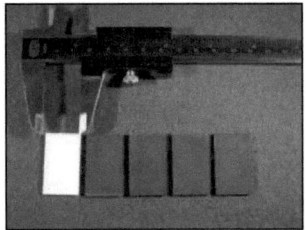

Fig. 4.5 Comparison between the as prepared EB-PVD PYSZ coating and the integrated coatings LaFeCo$_{(0.3)}$-Pd–PYSZ (from left to right).

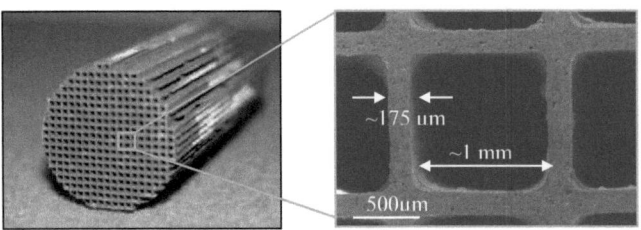

Fig. 4.6 Cordierite substrate coated with LaFeCo$_{(0.3)}$-Pd (left) and cross-section (right) air calcined at 700°C/3h.

Cordierite substrates provided by INTERKAT® were coated via the standard coating method employed at INTERKAT®. The particle size of the citrate-route synthesized powder of LaFeCo$_{(0.3)}$-Pd calcined at 500°C/3h was first reduced by milling in a ZrO$_2$ recipient with ZrO$_2$ balls and 2-propanol during 60 min. A final particle size was measured between 0.1 and 10 µm having a bimodal distribution. A water based suspension with 40 % of the solid powder perovskite and 5 % of a SiO$_2$ colloidal solution was prepared. The cordierite substrates were direct impregnated and the excess suspension inside the channels blew out with pressurized air. The same procedure was repeated two to three times. After the coating the catalyst mass on the cordierite substrate was 4.0102g. The sample was then dried with hot air at 150°C and calcined in oven up to 700°C for 3 h. The final mass of the catalyst on the ceramic monolith obtained after the drying and the calcination steps was 3.7218 g.

4.2 Characterization methods

A series of characterization methods are employed in order to analyze the synthesized catalyst. The next section describes these methods including the required sample preparation for each characterization method. The Differential Scanning Calorimetry (DSC) analysis was made in order to analyze the crystallization path in the catalysts. The X-Ray Diffraction (XRD) technique was employed to determine the crystal phase(s) present in the synthesized catalysts. The Rietveld analysis of the diffraction results provided an additional information about the crystal phase(s) purity of the catalyst and the redox behaviour of the Pd-substituted perovskites. The microstructure of the powder based catalysts and the integrated coating $LaFeCo_{(0.3)}$-Pd–PYSZ was investigated with a Field Emission Scanning Electron Microscope (FE-SEM). The catalyst composition and palladium state in the catalysts after the reduction and re-oxidation treatments was investigated with XRD, SEM and EDS. In addition to these characterization methods the catalyst's surface compositions were analysed under redox conditions via the XPS method by the research group of Dr. W. Grünert at the Technical Chemistry division of the Bochum University. Specific surface area determination of some of the perovskites was made via the BET method. The redox behaviour of the Pd-based catalysts was studied by means of XPS. NO_x reduction catalytic tests under lean conditions using H_2 as a reducing agent were carried out at CenTACat of the Queen's University in Belfast. During these experiments the analysis of the different reactants and products was achieved with a chemiluminescent analyzer for NO and NO_x measurement. A FT-IR spectrometer was employed for the gas analysis such as; CO, N_2O, CO_2 and H_2O and NH_3 only present under rich conditions (no oxygen). The use of propene (C_3H_6) as a reduction agent for NO_x reduction under lean conditions was proved in our facilities at the Institute of Materials Research of DLR. The analysis of the reactants and products during these experiments was carried out with the Mass Spectrometry.

4.2.1 Microstructure and composition analysis

Microstructural characteristics of the as-prepared powder catalysts were determined by means of the Field Emission-Scanning Electron Microscope (FE-SEM) LEO Gemini GSM 982 equipped with an Energy Dispersive X-Ray Spectrometer (EDS) from Oxford Instruments and the Transmission Electron Microscope from Phillips EM 430. The acceleration voltage for the EDS analysis was 15 kV. Small amount of powder sample was sieved to obtain a particle size of 63 µm and attached to a carbon adhesive tape. The weakly bonded agglomerates were blow out of the sample holder with dry air. The samples were coated with platinum in a Sputter Coater SCD 500 from BAL-TEC.

A careful metallographic preparation of the ceramic coatings is required in order to make a reliable microstructural analysis. The cross-sections of the integrated coatings $LaFeCo_{(0.3)}$-Pd–PYSZ were cut in small pieces and embedded in a conductive phenolic mounting compound with a press (LaboPress-1) from Struers applying 25 kN and 180°C for 10 to 15 min. The samples were then wet-polished with abrasive paper 500, 800, and 1200. Final polishing with a diamond spray and a colloidal SiO_2 suspension was carried out with an automated RotoPol-31 from Streuer applying 25 kN at 150 rpm during 3 minutes in each case. The samples were thoroughly cleaned with soap and plenty of water, and dried at 90°C in an oven for 30 minutes before analysis. A conductive platinum coating was sputtered on the samples with the coater equipment mentioned above. A X-Ray Fluorescence Analyzer MESA 5000 from Oxford Instruments was used for the catalyst composition determination.

XPS analysis of the catalysts

X-ray photoelectron spectroscopy also known as ESCA (Electron Spectroscopy for Chemical Analysis) is a surface chemical analysis technique that in our case was used to determine the surface composition and the electronic state of the elements of our catalysts. The photoelectron spectroscopy utilizes the photo-ionization and the energy dispersive analysis of the emitted photoelectrons to determine the composition and the electronic state of a determined surface region of a sample within the first 1 to 10 nm. The reason is the short inelastic mean free path (IMFP) of the generated photoelectrons. This means that emitting photoelectrons generated as a result of the x-ray will be penetrated 1-5 μm into the material and thus recaptured. The most commonly employed monochromatic x-ray sources are: the Mg K_α radiation $h\upsilon$ = 1253.6 eV and the Al K_α radiation $h\upsilon$ = 1486.6 eV. In the XPS the photon is adsorbed by an atom in a molecule or a solid leading to the ionization and the emission of a core electron (inner shell). The kinetic energy distribution of the emitted photo-electrons can be measured using an appropriate electron energy analyzer and a photoelectron can thus be recorded. The basic equation of the XPS method firstly reported by Rutherford in 1914 is;

$$E_K = h\upsilon - E_B$$

were E_K is the kinetic energy of the photo-electrons (β-rays), $h\upsilon$ is the incident photon energy and E_B is the electron binding energy [86]. The binding energy is taken to be a direct measure required to remove electrons from its initial level to the vacuum level.

Table 4.2 catalyst's formulations analyzed by means of XPS

Catalyst composition [a]	Treatment(s)
$LaFeCo_{(0.35)}$ [c]	As prepared
Pd-$LaFeCo_{(0.35)}$	As prepared
Pd-$LaFeCo_{(0.35)}$	200°C, in 4.2 % H_2 + Ar [b]
$LaFeCo_{(0.3)}$-**Pd**	As prepared
$LaFeCo_{(0.3)}$-**Pd**	200, 500°C in 4.2 % H_2 + Ar [b]
BaTi [c]	As prepared
Pd-BaTi	200, 500°C in 4.2 % H_2 + Ar [b]
BaTi-**Pd**	500°C in 4.2 % H_2 + Ar [b]

[a] Theoretical catalyst compositions are implied
[b] reduction for 2 h
[c] samples analyzed as received at room temperature

Table 4.2 summarises the catalyst's compositions analyzed with the XPS characterization method. The Pd-integrated perovskite (with palladium in the crystal lattice) $LaFeCo_{(0.3)}$-Pd were analyzed with XPS after the reduction treatments in 4.2 vol. % H_2 + Ar at 200° and 500°C (Table 4.2). The palladium supported perovskite Pd-$LaFeCo_{(0.35)}$ was only analyzed after the reduction at 200°C because the complete reduction of palladium was achieved at this temperature and the perovskite with palladium in the crystal lattice labelled as Pd-substituted perovskite BaTi-Pd was only analyzed after the reduction treatment at 500°C. The analysis of the BaTi-Pd perovskite after the reduction at 200°C was not carried out because of a poor spectral quality.

4.2.2 Crystal phase(s) analysis

The Differential Scanning Calorimetry (DSC) analysis was made with a NETZSCH 404 analyzer. Tablets of the $LaFeCo_{(0.3)}$-Pd-500°C/3h having a diameter of 4.65 mm were prepared with a manual press. The temperature was raised up to 900°C at a constant heating rate of 10°C/min under static air. The DSC experiments under the same conditions were carried out up to the detected transition temperatures and immediately cooled down to room temperature: After that the samples were analyzed with the XRD technique.

The X-Ray Diffraction (XRD) powder method was employed to determine and analyze the phase(s) of the synthesized samples in the as-prepared state and after the different reduction and re-oxidation treatments. After the catalytic test, the perovskite powders were also XRD-analyzed. The XRD diffraction patterns presented in this work were obtained by using a diffractometer of type SIEMENS D5000 in Bragg Brentano geometry using Cu $K\alpha_{1+2}$ radiations. The samples were slightly grounded in an agate mortar and mounted on Si single crystal sample holders just by adhesion using 2-propanol for dispersion. Peak profile analysis of the diffraction results was achieved with the program Diffract AT from Siemens 1993. Further analysis of the XRD results employing the Rietveld refinement was done for the lattice parameters and the phase(s) composition calculation employing the program Wyriet 3 [87].

Reduction and re-oxidation of the Pd substituted perovskites

In order to demonstrate the reversible segregation of the Pd ions out of the crystal lattice in the $LaFeCo_{(0.3)}$-Pd and BaTi-Pd based perovskites, powder samples of both catalyst compositions were reduced under 5 vol. % H_2 in N_2 employing an Al_2O_3 ceramic tube with an internal diameter of 2.5 cm. The reduction treatments of the synthesized perovskites were carried out with a constant flow of 50 ml.min^{-1} at 600°C for 1 h. During cooling of the sample a constant flow rate of the gas mixture was maintained until the oven reached approx. 250°C. Further cooling of the sample to room temperature was achieved under a constant flow of nitrogen to avoid further oxidation processes. The reduced Pd substituted perovskites were oxidized at 600°C for 3h in an oven under static air. Finally, the reduced and re-oxidized samples were analyzed by means of XRD, TEM and SEM.

4.2.3 Surface area(s) measurements

The Brunauer-Emmett-Teller (BET) method was employed to calculate the specific surface area of the catalysts. The method is an extension of the Langmuir theory, which is a theory for monolayer molecular adsorption on solid surfaces, to multilayer adsorption with the following considerations; (a) the gas molecules (i.e. N_2) physically adsorbed on a solid in layers infinitely, (b) there is no interaction between each adsorption layer, and (c) the Langmuir theory can be applied to each layer, the BET equation is expressed by (1),

$$\frac{1}{\upsilon[(P_0/P)-1]} = \frac{c-1}{\upsilon_m c}\left(\frac{P}{P_0}\right) + \frac{1}{\upsilon_m c} \qquad (1)$$

P and P_0 are the equilibrium and the saturation pressure of the adsorbates at the temperature of adsorption, υ is the adsorbed gas quantity (in volume units), and υ_m is the monolayer adsorbed gas quantity, c is the BET constant, which is expressed by (2),

$$c = \exp\frac{E_1 - E_L}{RT} \qquad (2)$$

E_1 is the heat of adsorption for the first layer, and E_L is that for the second and higher layers and is equal to the heat of condensation. Equation (1) is an adsorption isotherm and can be plotted as a straight line with $1 / v[(P_0 / P) - 1]$ on the y-axis and $\varphi = P / P_0$ on the x-axis according to the experimental results (The BET plot). The linear relationship of this equation is maintained only in the range of $0.05 < P / P_0 < 0.35$. The value of the slope A and the y-intercept I of the line are used to calculate the monolayer adsorbed gas quantity υ_m and the BET constant c. A total surface area S_{total} and a specific surface area S_{BET} can be calculated by the following equations;

$$S_{total} = \frac{\upsilon_m N s}{V} \qquad (3)$$

$$S_{BET} = \frac{S_{total}}{a} \qquad (4)$$

Where N is the Avogadro's number, s is the adsorption cross-section (for N_2 $s = 0.16$ nm^2), $V =$ molar volume of adsorbed gas (22.4 l.mol^{-1} under STP), and $a =$ molar weight of adsorbed species.

4.2.4 Determination of the catalytic activity

Two reactions were studied in this work; the NO_x-reduction under lean conditions using hydrogen as a reducing agent and the NO_x-reduction under lean conditions employing propene as a reducing agent. The H_2-SCR reaction was carried out at CenTACat (Centre of the Theory and Application of Catalysis) at Queen's University Belfast. The effects of CO_2 + H_2O_{vapour} and CO in the feed to the catalytic performance of the perovskite catalysts were studied. The second test reaction was carried out to prove the dry NO_x-reduction over the perovskite catalysts under lean conditions employing propene as a reducing agent (C_3H_6-SCR of NO_x). The NO_x reduction ability of the perovskite catalysts was investigated by changing the oxygen and the propene concentrations in the feed.

4.2.4.1 Selective Catalytic Reduction (SCR) of NO_x with H_2

Prior the test reaction each catalyst was pelletized. The discs having 1.5 cm of diameter were prepared by using a small amount of catalyst powder and applying a pressure of 9 tonnes during 30 - 45 seconds. After that, the samples were slightly crushed into small pieces with an agate mortar. The agglomerates were grounded, sieved and separated with mesh sizes 250 and 450 μm. 300 mg of fresh catalyst was fixed inside a quartz tubular reactor at atmospheric pressure (o. d. = 5 mm). The catalyst in each experiment was supported over a plug of quartz wool and maintained in a constant position using a hollow quartz support rod. All the lines were heated and thermal isolated to avoid water condensation. The temperature of the catalyst was measured by a thermocouple in direct contact with the bed catalyst. The NO and NO_2 signals were continuously monitored using a chemiluminescent NO/NO_x analyzer SIGNAL series 4000. Prior to the NO_x analyzer the effluent gas mixture was passed through a water trap. The gas phase was analyzed with an IR spectrometer from BRUKER. Fig. 4.7 displays the flow diagram of the reaction system. Gas mixtures with NO, H_2, O_2, CO_2 and H_2O including increasing concentrations of N_2O (all BOC gases) were prepared in order to calibrate and calculate the N_2O formed during the reaction (see appendix 3 for calibration curves). The integration of the area below the corresponding IR bands 2202 and 2236 cm^{-1} allows us to calculate the amount of N_2O present in the gas mixture. Since NH_3 was not detected during our experiments, the N_2 concentration was calculated by the difference between the feedstock and the effluent concentrations of the N-containing species.

The evaluation of the results was made by applying the following equations:
The NO_x-conversion to N_2 and N_2O

$$X = \frac{[NOx_{in}] - [NOx_{out}]}{[NOx_{in}]} * 100 \, (\%)$$

% Selectivity to N_2O
$$S_{N_2O} = \frac{2[N_2O]}{[NOx_{in} - NOx_{out}]} * 100 \, (\%)$$

-% Selectivity to N_2
$$S_{N_2} = \frac{[NOx_{in} - NOx_{out}] - 2[N_2O]}{[NOx_{in} - NOx_{out}]} * 100 \, (\%)$$

$W/F = 0.065 \, g_{cat}.s.ml^{-1}$

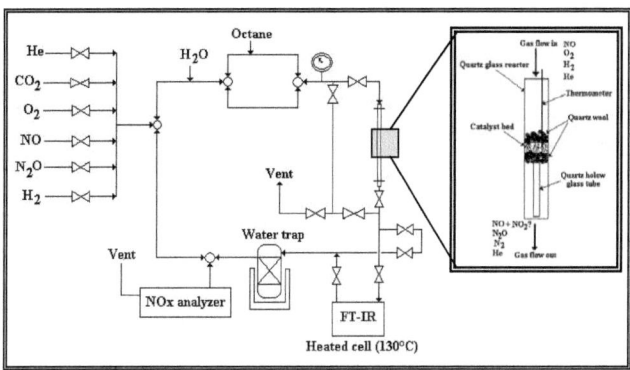

Figure 4.7 Schematic flow diagram of the reaction test unit at Queen's University CenTACat.

4.2.4.2 Selective Catalytic Reduction (SCR) of NO_x with propene

The powder catalyst(s) were pressed under a load of 8 tons for 30-45 sec and then slightly grounded in an agate mortar. The agglomerates were sieved and separated with the sieve sizes of 250 and 450 μm. Fig. 4.8 shows the microstructure and the size of the pelletized catalyst. 75 mg of fresh catalyst were fixed inside a vertical micro-reactor with two plugs of quartz wool during each catalytic test. A quartz tubular reactor having an outer diameter of 5 mm under atmospheric pressure was used. The temperature during the catalyst test was measured with a thermocouple located outside the reactor. The analysis of the educts and products was made with a Mass Spectrometer QMS 422 from Pfeiffer Vacuum. The zero background of the MS was made with a constant flow of He during 20 h. The O_2, C_3H_6, NO, CO_2, H_2O, and Ar were simultaneously detected and recorded on-line by mass spectrometry with the mass numbers reported in table 4.3. CO was not measured because of the mass overlapping with CO_2 which is formed after combustion of C_3H_6. The nitrogen formation was only qualitatively identified employing the masses 14 and 28. The apparent nitrogen concentration was strongly influenced by the presence of the other components, CO_2, NO, CO. Before each experiment a constant flow of 300 ml.min^{-1} of Ar was maintained for 1.5 h to eliminate the adsorbed gases (i.e. water) inside the vacuum chamber. The selected gas composition for the reaction was then prepared employing diluted gases in Ar prepared by Infra. The gas flow was controlled with mass flow controllers from MKS. All the gases were first introduced in a mixing chamber and then conducted to the micro-reactor. Argon was always used as a dilution gas. Each time a blind test was done under the same reaction conditions using only a plug of quartz wool. Figs. 4.9 and 4.10 show the test unit (SESAM) employed for the catalytic test located at the Institute of Materials Research of DLR. The SESAM unit can also be employed to characterize the sensing properties of gas sensors by means of the impedance spectroscopy [88].

Figure 4.8 Pelletized perovskite $LaFeCo_{(0.3)}$-Pd-700°C prior catalytic test.

Table 4.3 Mass numbers for analysis of gases during the NO_x reduction with C_3H_6.

Gas	Mass number(s)
Ar	20
N_2	14, 28
CO_2	12, 13, 22, 44, 45
C_3H_6	26, 37, 39
H_2O	17, 18
NO	30
O_2	32

The analysis of the results was made by applying the following formula:

The NO_x conversion:

$$X = \frac{[NOx_{in}] - [NOx_{out}]}{[NOx_{in}]} * 100 \, (\%)$$

$W/F = 0.015\ g_{cat}.s.ml^{-1}$

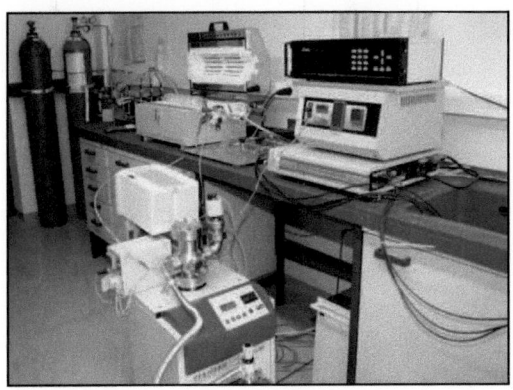

Figure 4.9 Sensor and catalyst test unit (SESAM), Research Institute of Materials at DLR

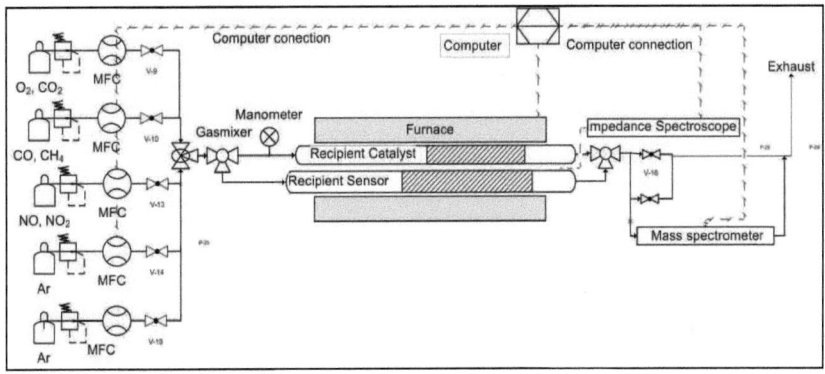

Figure 4.10 Schematic flow diagram of the sensor and catalyst test unit (SESAM) at DLR.

Chapter 5

5. Results

5.1 Microstructure and phase analysis of the powder catalysts

The following chapter presents the microstructure and phase determination of the lanthanum and barium based perovskite powder catalysts. In order to determine the crystal phase(s) evolution the catalysts were heat-treated at different calcination temperatures. It was mainly aimed to answer, if the synthesis methods employed in this work are suitable to incorporate precious metal ions (i.e. Pd, Rh) into the crystal lattice of the perovskite. Therefore, particular attention is being paid in this section to the state of palladium as a function of calcination temperatures of the substituted perovskite structures $LaFeCo_{(0.3)}$-Pd and BaTi-Pd. Redox behaviour of the Pd-substituted perovskites after being exposed to hydrogen rich atmospheres was studied by means of XRD, XPS, and SEM.

5.1.1 Microstructure and crystallization path of Pd-substituted perovskite ($LaFeCo_{(0.3)}$-Pd)

Fig. 5.1 shows the results obtained by the DSC measurement of the palladium substituted La-based perovskite which was heat-treated at 500°C/3h prior to the DSC measurements. The DSC spectrum shows two exothermic signals; one at 600°C with a weight loss of 3.2 mass % and the second at 640°C with a weight loss of 5.1 % due to crystallization of the sample. No additional mass loss is detected after calcination at 650°C. Fig 5.2 displays the crystal phase(s) evaluation of the perovskite catalyst $LaFeCo_{(0.3)}$-Pd prepared by the citrate route and calcined at different temperatures. The XRD patterns of the perovskite powder $LaFeCo_{(0.3)}$-Pd calcined at 500°C shows the typical characteristics of a highly amorphous material which begins to crystallize at around 560°C as seen with the DSC measurement given in Fig. 5.1.

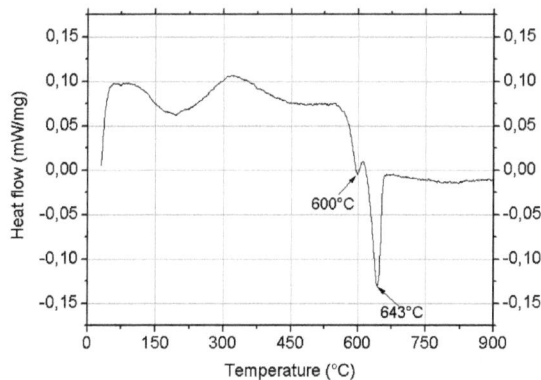

Figure 5.1 DSC measurement of the perovskite $LaFeCo_{(0.3)}$-Pd previously calcined in air at 500°C/3h

The early well-defined signals of the perovskite are observed at the XRD spectra of the Pd-substituted perovkite calcined at 600°C due to the start of its crystallization (Fig. 5.2). It is important to note that this spectrum still shows some characteristics of an amorphous phase as can be seen particularly at lower 2θ values. Fully crystallized perovskite is obtained at around 660°C. The diffractogram of the sample after DSC-measurement at 660°C also shows well-defined reflections corresponding to a single orthorhombic phase with the composition LaFeCo$_{(0.3)}$-Pd. At this temperature no other single or mixed oxides were detected and no reflections related to the presence of palladium oxide were found, suggesting the palladium substitution of the B-site in the perovskite structure (ABO$_3$).

After calcination of the perovskite LaFeCo$_{(0.3)}$-Pd at 700°C/3h the crystallinity increases, small signals related to traces of cobalt oxide (Co$_3$O$_4$) appear (see Fig. 5.3). The weak reflection at 2θ = 30.9° of the perovskite calcined at 700°C is related to the presence of this small amount of cobalt oxide according to data JCPDS 80-1543. The substitution with Pd at the expense of Co at the B-site in the perovskite LaFe$_{0.65}$Co$_{0.35-x}$Pd$_x$O$_3$ lead to lower cobalt content and less cobalt oxide segregation. At 750°C no significant changes in the main perovskite structure were observed, but segregation of Pd out of the crystal lattice as PdO is detected. At 900°C the crystallinity of the sample increases displaying the reflections of the orthorhombic perovskite structure and tetragonal PdO. Synthesis of the perovskite LaFeCo$_{(0.35)}$ (without palladium) under the same conditions helps us to prove this affirmation. As shown in Fig. 5.3 there are no signals at 2θ = 33.9° in the Pd-free perovskite after calcination at 900°C which indicates that no overlapping diffraction signals in the spectra of the perovskite and PdO exist.

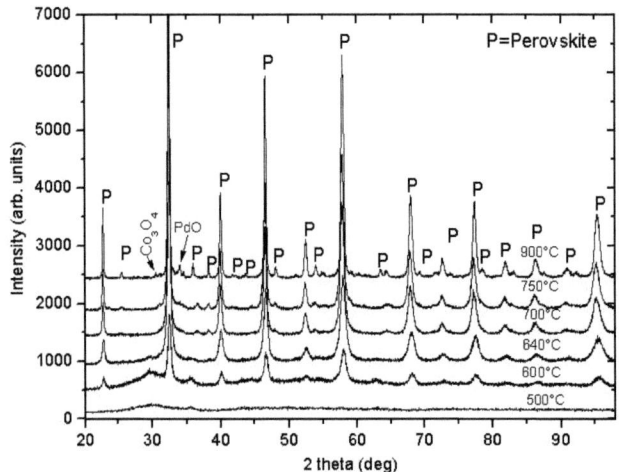

Figure 5.2 Crystal phase evolution of the catalyst LaFeCo$_{(0.3)}$-Pd calcined under static air.

The microstructure of the perovskite catalysts LaFeCo$_{(0.3)}$-Pd is presented on Fig. 5.4. A porous connected particulate appearance was obtained for a catalyst calcined at 900°C/3h (BET surface area = 1.657 m^2/g). The micrographs in Fig. 5.4 clearly show that the crystallite sizes of the synthesized perovskite are partially below 200 nm. The microstructure shows also a variety of pore geometries and sizes which reach in some cases up to 350 nm.

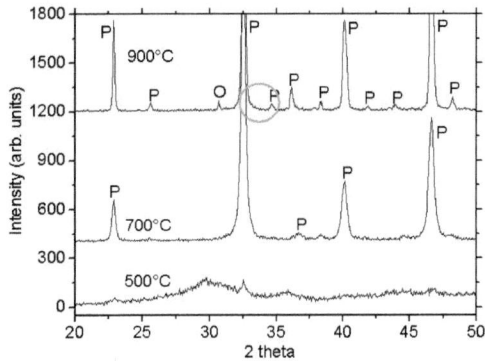

Figure 5.3 Crystal phase evolution of the perovskite LaFeCo$_{(0.35)}$ after calcination 500° - 900°C/3h under static air (P = Perovskite). The circle points out the 2θ angles were palladium oxide appears when present.

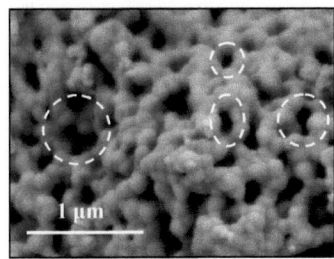 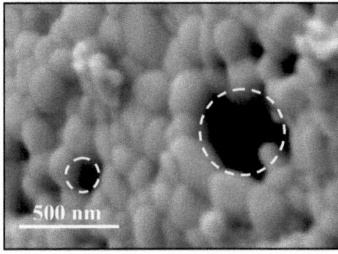

Figure 5.4 Microstructure of the perovskite LaFeCo$_{(0.3)}$-Pd after calcination at 900°C/3h at 30000 X (left) and 50000 X (right).

According to the data reported by McReady et al [89] the synthesis method plays a decisive role in the final perovskite crystal phase. They prepared the cubic perovskite phase with LaCo$_{0.4}$Fe$_{0.6}$O$_3$ composition by calcining amorphous citrate precursor at 700°C/2h and the orthorhombic perovskite phase with the same composition by the glycine-nitrate combustion process after calcination at 850°C/6h. Rietveld refinement analysis of the diffraction patterns indicate that the perovskite synthesized by the citrate route in this study is of orthorhombic perovskite structure and single crystal. Fig. 5.5 shows the Rietveld refinement of the XRD pattern of the perovskite LaFeCo$_{(0.3)}$-Pd-700°C including the Bragg positions of the corresponding orthorhombic and cubic perovskites with the same chemical composition. According to the Rietveld refinement analysis, the presence of cubic perovskite LaFeCo$_{(0.3)}$-Pd can be excluded. The Lowest Bragg positions in Fig. 5.5 correspond to tetragonal PdO. PdO was not detected by XRD on the perovskite LaFeCo$_{(0.3)}$-Pd after its calcination in static air at 700°C.

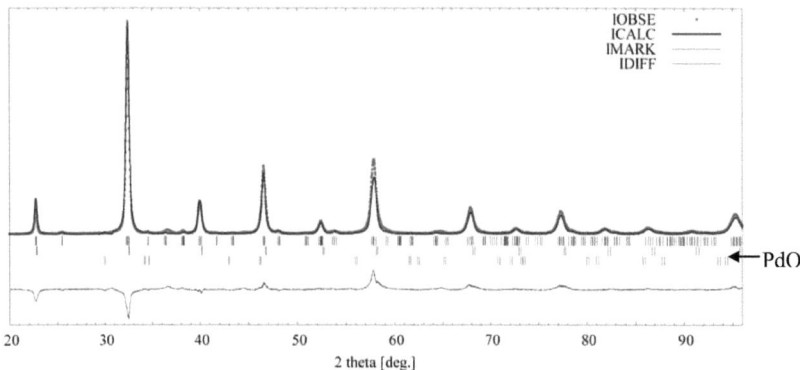

Figure 5.5 Rietveld refinement of the diffraction pattern from LaFeCo$_{(0.3)}$-Pd air calcined at 700°C/3h including orthorhombic (space group Pnma) and cubic perovskites (space group Pm3-m). The lowest Bragg positions in this figure correspond to PdO. Data collected from 2θ = 20 to 98°, step size = 0.02°, counting time = 20 sec/step, and 0.2 mm slit.

The EDX-analysis of the Pd-integrated perovskite (Fig. 5.6) showed a relatively homogeneous distribution of palladium in the catalyst (see Table 5.1). The 0.05 mol fraction corresponds to approx. 2.5 wt. % as PdO in the La-based perovskite. The presence of carbon in our EDX analysis is due to the carbon film employed to fix the powder sample. The acceleration voltage was 15kV and the beam diameter approx. 0.5 µm. It must be noted that due to the relatively wide electron beam diameter in the FE-SEM, the EDX-analysis is only a qualitative measurement. However, the SEM investigation and the EDX analysis of the Pd-substituted sample LaFeCo$_{(0.3)}$-Pd and the Pd-supported sample Pd-LaFeCo$_{(0.35)}$ give us insight about the Pd-state in both perovskites. The EDX-analysis of the LaFeCo$_{(0.3)}$-Pd suggests the homogenous distribution of Pd in the crystal structure and the catalyst. A high dispersion of palladium in the catalyst LaFeCo$_{(0.3)}$-Pd is assumed. In contrast to the Pd-substituted perovskite, the Pd-supported perovskite showed locally enriched concentration of palladium. Palladium is supported on the surface of a perovskite LaFeCo$_{(0.35)}$ which was previously air calcined at 900°C. Sponge-like agglomerates detected with the SEM are associated with the Pd-rich phase(s) in the Pd-supported perovskite Pd-LaFeCo$_{(0.35)}$ (see Fig. 5.7, points 1, 4 and 6).

Table 5.1 Palladium concentrations (wt. %) in the Pd-integrated perovskite LaFeCo$_{(0.3)}$-Pd calcined in air at 700°C/3h.

Spectrum	C	O	Fe	Co	Pd	La	Sum
1	3.91	27.76	14.74	7.02	2.33	44.24	100.00
2	7.89	28.76	13.51	4.89	2.32	42.62	100.00
3	4.78	25.50	13.82	7.38	-----*	48.52	100.00
4	3.94	23.56	15.80	6.30	2.23	48.17	100.00
Max.	7.89	28.76	15.80	7.38	2.33	48.52	
Min.	3.91	23.56	13.51	4.89	2.23	42.62	

*Pd was not detected in this point

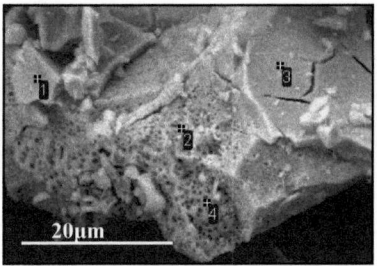

Figure 5.6 Pd-substituted perovskite LaFeCo$_{(0.3)}$-Pd-700°C as prepared; the numbers indicate the spectrum number of the EDX analysis listed in Table 5.1.

Figure 5.7 Pd-supported powder perovskite Pd-LaFeCo$_{(0.35)}$-700°C as prepared; the numbers indicate the spectrum number of the EDX analysis listed in Table 5.2.

Table 5.2 Palladium concentrations (in wt. %) after EDX- in the Pd-supported perovskite Pd-LaFeCo$_{(0.35)}$O$_3$ air calcined at 700°C/3h.

Spectrum	C	O	Fe	Co	Pd	La	Sum
1	4.88	19.84	12.44	7.84	11.22	43.78	100.00
2	5.02	21.80	14.77	7.27	------*	51.14	100.00
3	2.56	20.75	16.91	8.53	------*	51.25	100.00
4	7.09	20.30	9.82	5.99	19.89	36.92	100.00
5	3.74	21.49	14.32	8.38	------*	52.08	100.00
6	4.45	23.97	13.11	7.67	14.57	36.22	100.00
7	3.83	18.72	15.13	9.14	0.58	52.60	100.00
Max.	7.09	23.97	16.91	9.14	19.89	52.60	
Min.	2.56	18.72	9.82	5.99	0.58	36.22	

*Pd was not detected

a) Co-content at the B-site of the Pd-containing perovskite

Only lanthanum atoms are present at the A-site as three elements share the B-site including palladium in the perovskite based catalyst LaFeCo$_{(0.3)}$-Pd. The Fe/Co ratio was modified by increasing the Co-content and keeping the quantity of La and Pd constant. Higher contents of cobalt have a positive effect on the final properties of the catalyst (i.e. catalytic activity, N$_2$-selectivity, etc), as reported in section 5.3. The perovskite structure LaFe$_x$Co$_{0.95-x}$Pd$_{0.05}$O$_3$ (x= 0.475, 0.55, 0.65) is apparently composed of cubic and orthorhombic crystal phases (Fig 5.8) at calcination temperatures 700°C and lower. As the cobalt content increases in the perovskites LaFe$_x$Co$_{0.95-}$

$_x$Pd$_{0.05}$O$_3$, higher cobalt oxide (Co$_3$O$_4$) contents were detected on calcination at 900°C. The sharp increase of the reflection at 2θ = 31.1° and the appearance of the reflection at 2θ = 18.9° indicate that the cobalt excess in the catalysts segregates to form cobalt oxide (Co$_3$O$_4$) (see data JCPDS 80-1543). The increase of cobalt at the expense of iron at the B-site of the perovskite LaFe$_x$Co$_{0.95-x}$Pd$_{0.05}$O$_3$ (x = 0.65, 0.55, 0.475) causes a shift to higher 2θ values (see Fig. 5.9). This proves the increase of phase stability of the orthorhombic structure and the incorporation of cobalt to certain content into the main perovskite lattice.

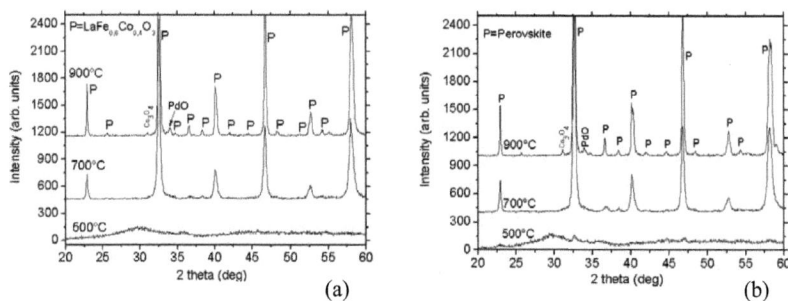

Figure 5.8 XRD powder patterns of the perovskites LaFeCo$_{(0.4)}$-Pd (a), and LaFeCo$_{(0.475)}$-Pd (b), calcined in air from 500 to 900°C/3h.

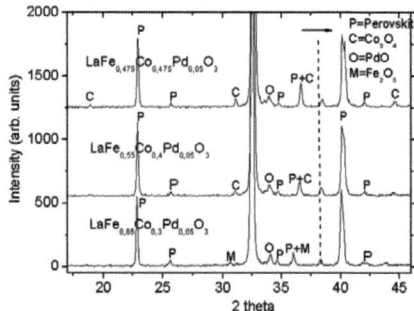

Figure 5.9 Evaluation of crystal phase(s) with increasing cobalt content present in the perovskite LaFe$_x$Co$_{0.95-x}$Pd$_{0.05}$O$_3$ (x = 0.65, 0.475, 0.4) calcined in air at 900°C/3h.

b) The cerium content in the LaCe$_{(0.05)}$FeCo$_{(0.3)}$-Pd perovskite

With the citrate method employed in the present work a highly homogenous perovskite with composition LaCe$_{(0.05)}$FeCo$_{(0.3)}$-Pd was obtained after calcination in air at 700°C for 3 h. No other phase(s) such as oxides were observed at this calcination temperature. After the perovskite exposure in air at 900°C/3h cerium and palladium segregate as oxides (Fig. 5.10). In addition a small amount of rhombohedral Fe$_2$O$_3$ was found (see data JCPDS 87-1166). Similar crystallization path was observed for the cobalt free perovskite LaCe$_{(0.05)}$Fe-Pd (see left graphic in Fig. 5.11). No crystallographic phase changes were observed at the lanthanum perovskite LaCe$_{(0.1)}$Fe-Pd after calcination at 700°C/3h. After calcination at 900°C/3h the perovskite LaCe$_{(0.1)}$Fe-Pd showed XRD signals of cubic cerium oxide (see data JCPDS 34-0394) that are more intense than the ceria XRD signals in the perovskite LaCe$_{0.05}$Fe-Pd implying a higher cerium oxide segregation with increasing cerium content in the catalyst (see right graphic in Fig. 5.11).

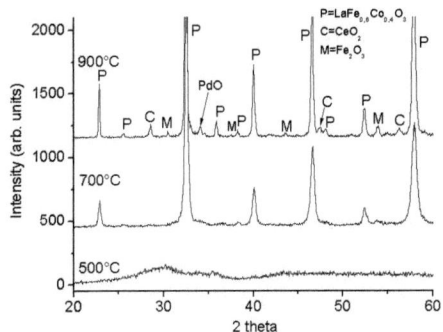

Figure 5.10 XRD patterns of the catalyst LaCe$_{(0.05)}$FeCo$_{(0.3)}$-Pd calcined in air.

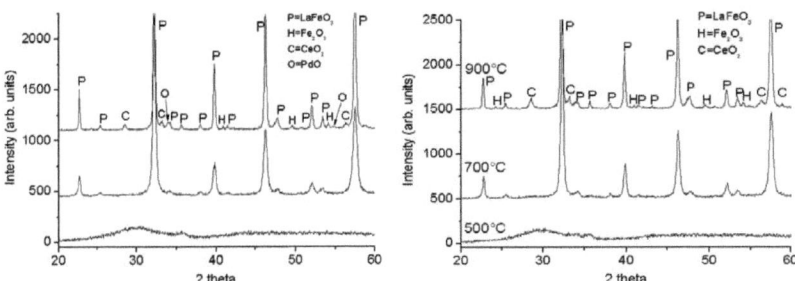

Figure 5.11 XRD patterns of the catalysts LaCe$_{(0.05)}$Fe-Pd (left) and LaCe$_{(0.1)}$Fe-Pd (right) calcined in air from 500°C up to 900°C.

High cerium loading in the perovskite LaCe$_{(0.4)}$Fe-Pd delayed its crystallization. After calcination at 700°C the sample was practically amorphous. The perovskite begins to crystallize at this temperature. On the other hand the relatively broad XRD buckle between 2θ = 28-31° implies the presence of high amount of amorphous ceria at 700°C. Well defined XRD reflections corresponding to cubic ceria are observed after calcination of the perovskite LaCe$_{(0.4)}$Fe-Pd in air at 900°C/3h (see Fig. 5.12).

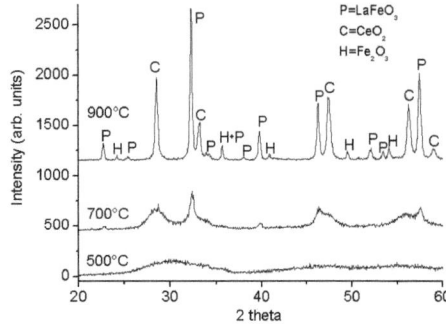

Figure 5.12 XRD patterns of the perovskite $LaCe_{(0.4)}Fe$-Pd calcined in air.

5.1.1.1 Redox behaviour of the $LaFeCo_{(0.3)}$-Pd perovskite

The Pd-substituted perovskite $LaFeCo_{(0.3)}$-Pd seems to be metastable up to 700°C, after this temperature PdO segregates out of the crystal lattice during calcination under oxidizing atmosphere in static air. According to the XRD results reported in the previous section it is assumed that palladium partially substitutes the B-site of the perovskitic crystal lattice calcined at 700°C. To study the redox behaviour of the perovskite, which was previously calcined in air at 700°C, some powder sample of $LaFeCo_{(0.3)}$-Pd is reduced under hydrogen rich atmosphere (5 vol. % $H_2 + N_2$) at 600°C/1h. Afterwards this reduced Pd- substituted perovskite is re-oxidized at 600°C for 3h in furnace under static air. Fig. 5.13 compares the phase changes occurred by reduction and re-oxidation of the perovskite $LaFeCo_{(0.3)}$-Pd. Slight structural changes can be observed after reduction of the perovskite as shown in the measured and calculated crystal lattice data given in Table 5.3. The diffusion of Pd ions out of the crystal structure is possibly related to these modifications. The lattice constants after reduction and re-oxidation treatments lie near, and thus, imply in fact a non-reversible behaviour in the La- perovskite structure. It must be noted that the reduction of the sample was carried out under very strongly reducing conditions under flowing gas and the re-oxidation of the perovskite was made in furnace under static air conditions. It is worth to note that the catalyst retains its main perovskite structure through the oxidation, reduction, and re-oxidation cycles. The perovskite structure of the catalyst $LaFeCo_{(0.3)}$-Pd was stable even after the reduction treatment, and no decomposition into several phases has been observed in this study.

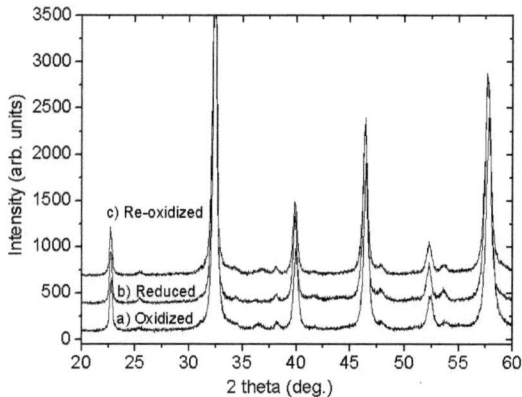

Figure 5.13 Powder XRD of LaFeCo$_{(0.3)}$-Pd ; (a) calcined at 700°C/3h, (b) after reduction in 5 vol. %H$_2$ + N$_2$ at 600°C/1h, (c) after re-oxidation at 600°C/3h in static air. Data collection from 2θ = 20° to 98°, step size = 0.02°, counting time = 20 sec/step and 0.2 mm slit.

Table 5.3 Lattice constants of the perovskites LaFeCo$_{(0.35)}$ and LaFeCo$_{(0.3)}$-Pd after redox treatments calculated by Rietveld refinement[*].

Sample	Treatment	a1	a2	a3	Phase(s)
LaFeCo$_{(0.35)}$-700°C	As prepared, calcined at 700°C	5,4998	5,5339	7,7926	Perovskite
LaFeCo$_{(0.3)}$-Pd-700°C	As prepared, calcined at 700°C	5,5024	5,5358	7,7972	Perovskite
LaFeCo$_{(0.3)}$-Pd-700°C	5%H$_2$ in N$_2$ calcined at 600°C	5,5262	5,5520	7,8093	Perovskite
LaFeCo$_{(0.3)}$-Pd-700°C	Re-oxidized in static air at 600°C	5,5214	5,5488	7,8052	Perovskite

[*] see appendix 1.

As-prepared Pd-substituted perovskite LaFeCo$_{(0.3)}$-Pd heat-treated at 700°C under oxidizing conditions showed a homogenous distribution of palladium as reported by EDX. SEM analysis of the as-prepared sample displayed no other phase(s) that could be related to the presence of segregated PdO or other oxide species such as; cobalt oxides and/or iron oxides (see micrograph (a) in Fig. 5.14). After reduction treatment of Pd-substituted perovskite LaFeCo$_{(0.3)}$-Pd in 5 vol. % H$_2$/N$_2$ at 600°C, some bright spots of very small size are observed. The presence of these nano-particles may be associated with the diffusion of palladium ions that form metallic palladium (micrograph (b) in Fig. 5.14). Unfortunately the diffraction peaks of metallic palladium and the perovskite LaFeCo$_{(0.35)}$ in the catalyst LaFeCo$_{(0.3)}$-Pd overlap, as a consequence metallic palladium cannot be identified by XRD. However, the bright spots disappear after catalyst re-oxidized at 600°C in air (micrograph (b) in Fig. 5.14) confirming the assumption made above as to the metallic palladium. The re-oxidation of LaFeCo$_{(0.3)}$-Pd, previously reduced in hydrogen rich atmosphere, may indicate the reversible diffusion of palladium into the crystal lattice of this perovskite.

In contrast to the Pd-substituted perovskite LaFeCo$_{(0.3)}$-Pd, the Pd-supported perovskite Pd-LaFeCo$_{(0.35)}$ shows clearly the presence of two phases after reduction treatment at 600°C.

Agglomerates of 50 nm and larger sizes are observed in the Pd-supported catalyst (see Fig. 5.15). These agglomerates are associated with the presence of metallic palladium.

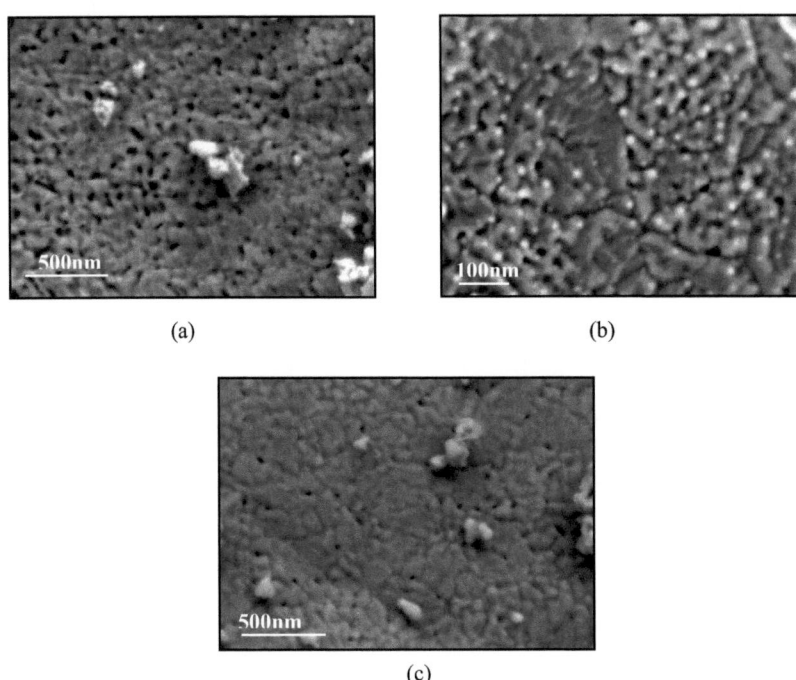

Figure 5.14 Comparison of the SEM-images of LaFeCo$_{(0.3)}$-Pd; calcined in air at 700°C/3h, 50000 X (a), after reduction treatment 5 vol. % H$_2$/N$_2$ at 600°C/1h, 150 000 X (b), and after re-oxidation in air at 600°C/3h, 50000 X (c).

Figure 5.15 Microstructure of Pd-LaFeCo$_{(0.35)}$ reduced in 50 ml.min^{-1} of 5 vol. % H$_2$/N$_2$ at 600°C/1h, 50000 X (a) and 100000 X (b).

Powder particles of LaFeCo$_{(0.3)}$-Pd calcined at 700°C were supported in cooper nets and analyzed by TEM. Evidence that indicates the presence of palladium particles as PdO on this perovskite after calcination under static air was found (see EDX of the crystal Kr1 in Fig. 5.16). However the presence of palladium in the matrix of the Pd-substituted perovskite LaFeCo$_{(0.3)}$-Pd state was also found in the oxidized powder sample (see EDX of the crystal Kr2 in Fig. 5.16). It seems like palladium or at least part of it may be still incorporated into the crystal lattice. Fig. 5.17 shows the agglomerated and particulate microstructure of the Pd-substituted perovskite LaFeCo$_{(0.3)}$-Pd-700°C in the oxidized state. In some cases no palladium particles were found in the perovskite in the as prepared state (oxidized state). On the contrary palladium in the matrix of the perovskite was detected (see EDX in Fig. 5.18).

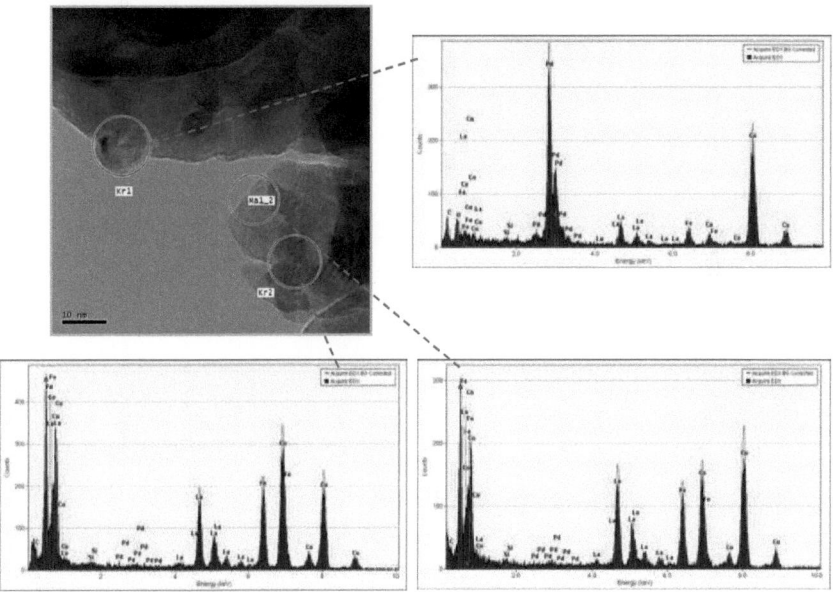

Figure 5.16 TEM-image and EDX analysis of the perovskite LaFeCo$_{(0.3)}$-Pd calcined in air at 700°C/3h.

Metallic palladium apparently diffuses out of the perovskite crystal lattice on reduction treatment at 600°C in 5 vol. % H$_2$/N$_2$ for 1h (Fig. 5.19). Bright spots on the TEM-image are related to metallic palladium formed probably as a result of palladium ions diffusion out of the crystal lattice. Palladium-rich particles (agglomerates) with sizes below 15 nm are observed. The upper EDX-spectrum suggests that these bright areas are Pd-rich zones thus supporting the assumption of palladium diffusion to form Pd°. The EDX-analysis taken at the darker areas (taken as a reference) showed the perovskite elements (i.e. La. Fe, Co), and no Pd. The TEM images in Fig. 5.20 show that intra-crystalline palladium (bulk-palladium) may diffuse out of the crystal structure on exposure to hydrogen containing atmospheres. The TEM-image on the right side in Fig. 5.20 displays a Pd-rich particle of 5 nm size proved after EDX analysis.

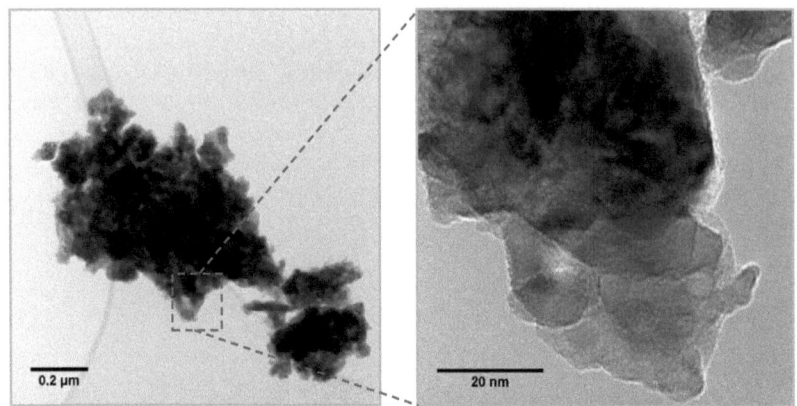

Figure 5.17 TEM-images of the perovskite LaFeCo$_{(0.3)}$-Pd calcined in air at 700°C/3h.

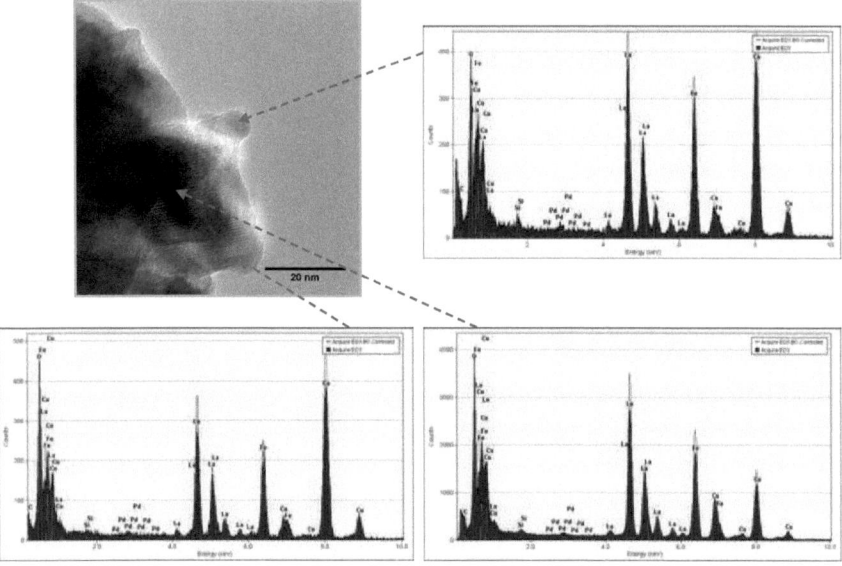

Figure 5.18 TEM-image and EDX analysis of the perovskite LaFeCo$_{(0.3)}$-Pd calcined in air at 700°C/3h.

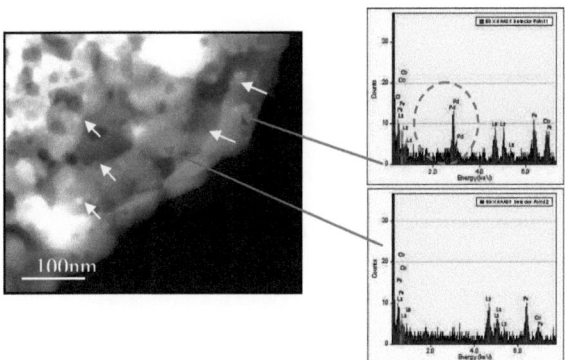

Figure 5.19 TEM images and EDX analysis of the perovskite LaFeCo$_{(0.3)}$-Pd after reduction treatment in 5 vol. % H$_2$/N$_2$ at 600°C 1h.

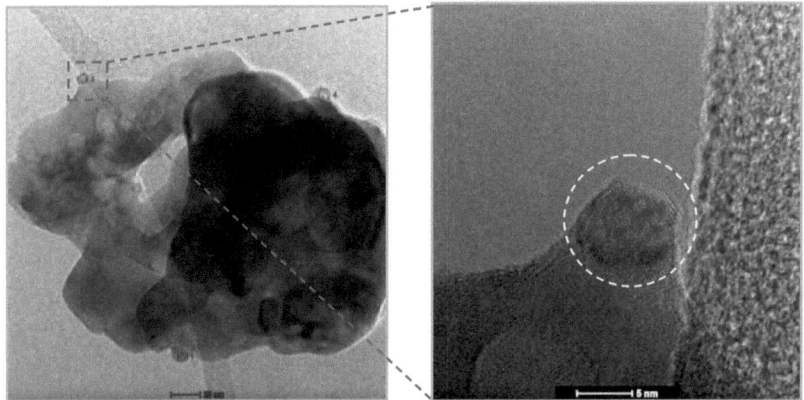

Figure 5.20 TEM images of perovskite LaFeCo$_{(0.3)}$-Pd after reduction treatment in 5 vol. % H$_2$/N$_2$ at 600°C 1h.

5.1.2 Microstructure and crystal phase(s) of Pd-substituted perovskites BaTi-Pd

In this work, a Pd-substituted perovskite having Ba and Ti at A- and B-sites BaTi-Pd was prepared via the co-precipitation method due to the nature of the precursors (see experimental part for detailed description). Since the degree of Pd-substitution in the Ba- based perovskite is unknown, the Pd-substituted perovskite is described by the formula BaTi-Pd On synthesis and heat-treatment of the Pd substituted catalysts at 500°C, a well crystallized perovskite phase with apparently tetragonal structure BaTiO$_3$ (BaTi) was obtained (Fig. 5.21 (b), see JCPDS data no. 75-0212). Depending on the transition temperature, BaTiO$_3$ presents different crystal structures. The stable phase at room temperature is the tetragonal phase. But some controversy arise when the crystallite size of BaTiO$_3$ decreases, the cubic BaTiO$_3$ phase can exist at room temperature suggesting that the phase transformation may be a function of temperature and the crystallite size [90]. In this study, the mathematical fit with a pseudo-Voigt function of the diffraction profile of the

BaTi-Pd perovskite assuming the presence of only the tetragonal phase yields satisfactory fits with the diffraction profile.

No peaks belonging to BaO or TiO_2 were detected by XRD. However, $BaCO_3$ with orthorhombic structure (JCPDS data no. 05-0378) was found for catalysts calcined at 500° and 700°C. It should be noted that due to the intensity of the diffraction peak at about $2\theta = 34°$, a considerable amount of $BaCO_3$ can be expected particularly at 500°C. $BaCO_3$ may decompose delivering CO_2 after air calcination at 700°C (Fig. 5.21 (a)). The diffraction peaks of the Pd-substituted perovskite BaTi-Pd-900°C located at $2\theta = 33.9°$, 31.1°, 42.1° and 61.1° correspond to probably small amounts of tetragonal palladium oxide (see JCPDS data no. 85-0624). Particularly at lower calcination temperatures of the catalyst, overlapping of the diffraction peaks of $BaCO_3$ and PdO hinders the clear identification of these two phases.

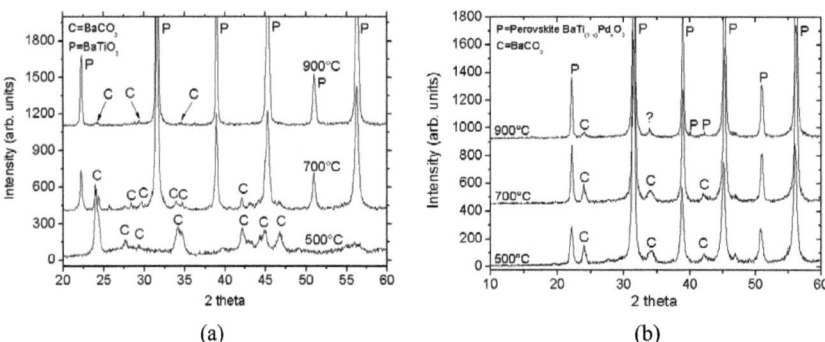

Figure 5.21 XRD patterns of the perovskites (a) $BaTiO_3$, (BaTi) and (b) BaTi-Pd calcined under static air from 500°C to 900°C/3h.

$BaCO_3$ is present in BaTi after calcination temperatures between 500°C and 700°C, as mentioned above. Traces of $BaCO_3$ are still observed on the Pd-free perovskite BaTi even after calcination of the catalyst at 900°C (see Fig. 5.21 (a)). Absence of PdO in the catalyst would be an indirect way to find out if the solid solution BaTi-Pd was formed. Due to overlapping in the Pd-substituted perovskite between the diffraction peaks of PdO and $BaCO_3$ it is not clear if Pd ions are completely or partially integrated in the perovskitic structure. The relative intensities of the peaks ($2\theta = 33.8°$) in the spectra at 900°C (diagram (b) Fig. 5.21) suggest the presence of PdO as a second phase. On the other hand the formation of BaTi-Pd as a solid solution at calcination temperatures 700°C and below is highly possible. If the solid solution BaTi-Pd was formed at temperatures below 700°C then the presence of PdO at higher temperatures would indicate that the structure is meta-stable. At higher temperatures the meta-stable perovskite structure would tend to reach its thermodynamic equilibrium causing phase(s) split. To support this idea Pd supported catalyst BaTi was prepared via the direct impregnation method. For this purpose perovskite BaTi was firstly calcined up to 900°C to completely eliminate $BaCO_3$ and avoid overlapping of the diffraction peaks from PdO and $BaCO_3$. Palladium content in the Pd-supported catalyst was maintained constant as in the Pd-substituted catalyst. Diffraction pattern of Pd-supported perovskite (Pd-BaTi-900°C)-700°C clearly shows the presence of PdO ($2\theta = 33.8°$), which is confirmed by the absence of diffraction peaks at the same 2θ Bragg angle in the perovskite BaTi-900°C (Fig. 5.22). Both samples; the Pd-supported Pd-BaTi catalyst and the Pd-free BaTi perovskite still have negligible amounts of $BaCO_3$ that do not interfere with the diffraction peaks of PdO (Fig. 5.22).

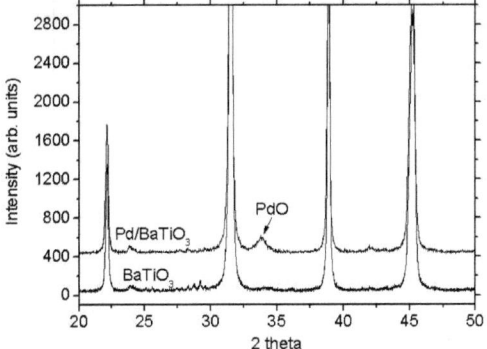

Figure 5.22 XRD spectra of the perovskites BaTi calcined at 900°C/3h and Pd-BaTi calcined in air at 700°C/3h.

Fig. 5.23 displays the porous microstructure of the Pd-substituted perovskite BaTi-Pd and the Pd-supported catalyst Pd-BaTi after calcination at 700°C/3h, both with approx. the same Pd-content ~2.5 wt. %. The Pd-substituted catalyst BaTi-Pd showed agglomerates with a relative homogeneous distribution of crystallites having sizes less than 100 nm. The synthesis method employed to prepare these samples explain the fine crystallite sizes observed. Soft elimination of the solvent (water) via the rota-vapour method minimizes the shrinkage of the gel network. Ba-based catalysts with higher BET surface areas than the La-based perovskites are obtained. The catalyst BaTi-Pd calcined at 500°C has a BET surface area of 45.1 m^2/g which decreases to 21.3 m^2/g after calcination at 900°C. However, compared with the state of the art $BaTiO_3$ nano-powders synthesized by soft chemistry method, the BaTi-Pd powders synthesized in this study showed remarkably higher BET surface areas. $BaTiO_3$ nano-powders calcined at 600°C with ~ 25 m^2/g have been reported in the literature [91].

Figure 5.23 Microstructure of Pd-substituted perovskite BaTi-Pd-700°C (a), and Pd-supported perovskite (Pd-BaTi-900°C)-700°C (b), both calcined in static air (30000 X).

5.1.2.1 Redox behaviour of the BaTi-Pd perovskite

In order to study the reversible redox behaviour of the catalyst, reduction of the Pd-substituted perovskite BaTi-Pd-700°C was carried out at 600°C under 5 vol. % H_2 containing N_2-atmospheres. Fig. 5.24 displays the microstructure of the Pd-substituted perovskite BaTi-Pd before and after reduction treatment. A highly porous microstructure is observed but no clear evidence of palladium diffusion out of the crystal structure or palladium nanoparticles formation was found. Palladium nanoparticles in the Pd-substituted perovskite may be a few orders of magnitude smaller than the palladium particles or agglomerates in the Pd-supported perovskite Pd-BaTi. Fig. 5.25 shows the microstructure of the palladium supported catalyst Pd-BaTi after reduction at 600°C for 1 h. High concentration of small size agglomerates or particles associated with metallic palladium are observed. These images imply that metallic palladium with extremely small size may be present in the Pd-substituted perovskite BaTi-Pd.

Figure 5.24 Microstructure of Pd-substituted perovskite BaTi-Pd as prepared and calcined in air at 700°C/3h (a), and after reduction with 5 vol. % H_2 + N_2 at 600°C/1h (b), (50000 X).

Figure 5.25 Microstructure of Pd-supported perovskite Pd-BaTi after reduction with 5 vol. % H_2 + N_2 600°C/1h; 30000 X (a) and 50000 X (b).

The XRD patterns for the catalyst BaTi-Pd before and after reduction treatment are compared in Fig. 5.26. Peaks associated with the presence of orthorhombic $BaCO_3$ (see data JCPDS 05-0378) are clearly visible even after reduction of the catalyst. After reduction in 5 vol. % H_2 + N_2 at 600°C for 1 h, metallic palladium with cubic structure was found (see data JCPDS 46-1043).

Metallic palladium disappears on re-oxidation at 600°C in static air for 1h. These results suggest that the Pd-substituted perovskite BaTi-Pd which may pose the crystal memory effect on redox conditions as mentioned in the literature [74].

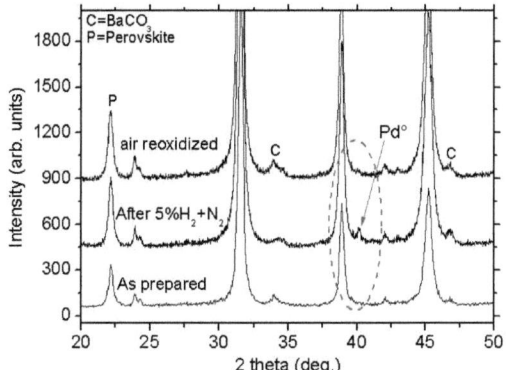

Figure 5.26 XRD pattern of the Pd-substituted perovskite BaTi-Pd calcined in air at 700°C/3h, after reduction in 5 vol. % H_2/N_2, at 600°C, and re-oxidized in air at 600°C/1h.

5.1.3 XPS analysis of the La- & Ba- based perovskite catalysts

In this work a BaTi perovskite is proposed as an alternative host to stabilize palladium ions. The perovskites with compositions Pd-BaTi, BaTi-Pd were analyzed with XPS as well. The XPS spectroscopy provides information about the chemical composition at the most outer surface layer of the catalyst. The XPS analysis may contribute to understand the redox behaviour of palladium-supported and -substituted perovskites.

Table 5.4 summarises the perovskite compositions analyzed by means of XPS. Table displays the sample condition treatment(s) before the XPS analysis. The Pd-free perovskite with composition LaFeCo$_{(0.35)}$ in as-prepared conditions was analyzed by means of XPS as reference. Similarly, the perovskite Pd-LaFeCo$_{(0.35)}$ as-prepared and after reduction treatment were analyzed. Palladium substituted perovskites LaFeCo$_{(0.3)}$-Pd and BaTi-Pd were measured by XPS after exposure to different reduction conditions at 200 and 500°C. The purpose of these experiments was to analyze the reduction behaviour of palladium in the catalysts Pd-LaFeCo$_{(0.35)}$ and LaFeCo$_{(0.3)}$-Pd.

Table 5.4 Perovskite compositions analyzed by means of XPS *.

No	Sample	Treatment before XPS measurement
1	LaFeCo$_{(0.35)}$	As prepared
2	Pd-LaFeCo$_{(0.35)}$	As prepared
3	Pd-LaFeCo$_{(0.35)}$	After red. at 200°C/2h in 4.2 % H$_2$ in Ar
4	LaFeCo$_{(0.3)}$-Pd	As prepared
5	LaFeCo$_{(0.3)}$-Pd	After red. at 200°C/2h in 4.2 % H$_2$ in Ar
6	LaFeCo$_{(0.3)}$-Pd	After red. at 500°C/2h in 4.2 % H$_2$ in Ar
7	BaTi	As prepared
8	Pd-BaTi	As prepared
9	Pd-BaTi	After red. at 200°C/2h in 4.2 % H$_2$ in Ar
10	Pd-BaTi	After red. at 500°C/2h in 4.2 % H$_2$ in Ar
11	BaTi-Pd	As prepared
12	BaTi-Pd	After red. at 500°C/2h in 4.2 % H$_2$ in Ar

*All samples calcined in air at 700°C/3h in the as prepared state.

Redox behaviour of the perovskite LaFeCo$_{(0.3)}$-Pd studied by means of XPS

Fig. 5.27 displays the XPS lines of all the elements of the Pd-free perovskite LaFeCo$_{(0.35)}$ heat treated in air at 700°C. The binding energy values of La 3d$_{5/2}$ indicate the trivalent form of lanthanum ions in the perovskite. The La 3d lines are characterized by very intense satellites (Fig. 5.27, a). The binding energy (BE) of La 3d$_{5/2}$ (~ 833.7 eV) is in good agreement with data previously reported in the literature [80]. These binding energies are pretty similar to those of LaCoO$_3$ measured by the group of Prof. W. Grünert [92]. The binding energy of La 3d$_{5/2}$ was 833.1eV in LaCoO$_3$ and the signal shape was identical with that shown here.

The oxides of Fe^{2+} and Fe^{3+} have different binding energies in their oxides (BE FeO = 709.5 - 709.9 eV, BE Fe$_2$O$_3$ = 711eV), but it is likely that binding energies are strongly influenced by changes in chemical bond (covalent vs. ionic) and by intra-crystalline potentials (Madelung term). Therefore, satellites are better tools to identify oxidation states. In Fe(III) oxidation states, the satellite is detached from the main line, appearing at BE 8.5 eV higher BE, in Fe(II) oxidation states, its BE is only 6 eV higher than that of the main line. Fig. 5.27 b shows the weak satellite line of iron, which is in the trivalent state as expected and was observed in all Fe-spectra, even after reduction of the Pd-substituted perovskite LaFeCo$_{(0.3)}$-Pd at 500 °C. As a partial conclusion, iron (Fe^{3+}) was not reduced after the pretreatment in H$_2$.

The BE is not a good parameter to distinguish between the bi- and trivalent states due to the small difference (binding energies between 780.7 and 781.3 given for CoO, Co$_3$O$_4$ and Co$_2$O$_3$ in the NIST data base (BE of Co 2p$_{3/2}$ in LaCoO$_3$ = 779.8 eV) [92]. In this case the satellites are better, because Co^{2+} has an intense satellite whereas that of Co^{3+} is weak. Moreover, the (apparent) spin-orbit splitting is at approx. 15 eV for Co^{3+} and 15.7 eV or higher for Co^{2+} (see Table 5.5). This results agree with spin-orbit splitting values reported in the literature [80] and references therein. The satellite in the perovskite LaFeCo$_{(0.35)}$ is much stronger than in LaCoO$_3$ (see inset in Fig. 5.27 c), which indicates that a significant part of cobalt in the surface should be in the bivalent state.

There are generally two O1s states, one around a BE of 529 eV (lattice oxygen) and one at 530 - 531 eV associated with OH groups and/or carbonates (Fig. 5.27 d). Eventually, the latter state may arise from oxidized species on the residual carbon at the catalyst surface. The ratio between lattice O and the other state is in around 2/1. In some samples there is also a very weak state round 526 eV, which is quite untypical for oxygen.

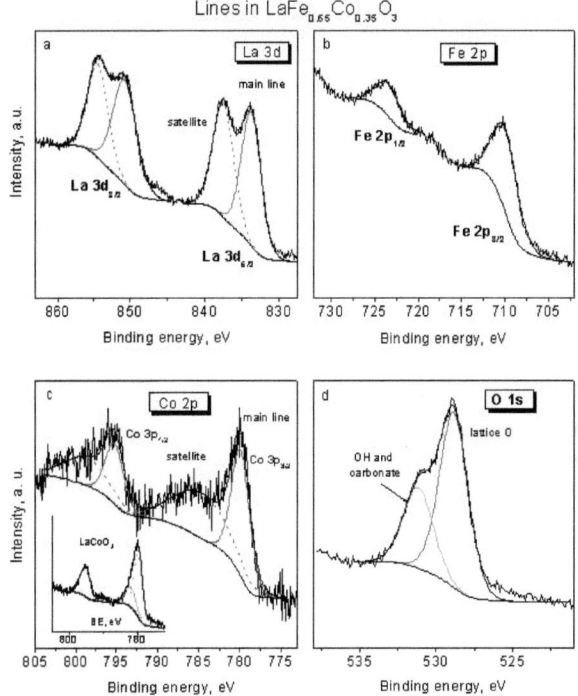

Figure 5.27 XPS lines of the Pd-free perovskite $LaFeCo_{(0.35)}$ calcined in air at 700°C/3h.

According to the satellite-main line ratio, after palladium impregnation [Pd-LaFeCo$_{(0.35)}$] the oxidation state of cobalt should be still near 3+ (see Table 5.15). On the other hand, according to the spin-orbit splitting, it should be Co^{2+}. Closer inspection reveals that the spin-orbit splitting could be everywhere between 15.4 and 16 eV in this sample indicating that the cobalt oxidation state may not be changed after impregnation of the perovskite LaFeCo$_{(0.35)}$ with palladium. A spin-orbit splitting value for CoO was reported as 15.5 eV or 16 eV [80] and references therein.

In LaFeCo$_{(0.3)}$-Pd, the average oxidation state of cobalt is clearly closer to Co^{2+} than in the Pd-free sample – both satellite-main line ratio and spin-orbit splitting are larger than in the latter. To make sure that this is not a fitting artifact; the two lines were compared after satellite subtraction and normalization to the peak height of the Co $2p_{3/2}$ component [Fig. 5.28 (a)]. In this plot, in which the Co $2p_{1/2}$ region should be disregarded due to the larger influence of baseline problems, an increased satellite contribution in the Pd-containing sample is obvious [Fig. 5.28 (a)]. This is a clear indication that the whole perovskite lattice has been perturbed, i.e., a proof for (at least partially) palladium integration into the crystal lattice. The fitting results after R200 and R500 indicate a further increased contribution of Co^{2+} in the reduced samples. According to the comparison made in

Fig. 5.28 (b), reduction at 200°C caused an increase of the satellite intensity whereas reduction at 500°C leaves the cobalt oxidation state essentially unchanged.

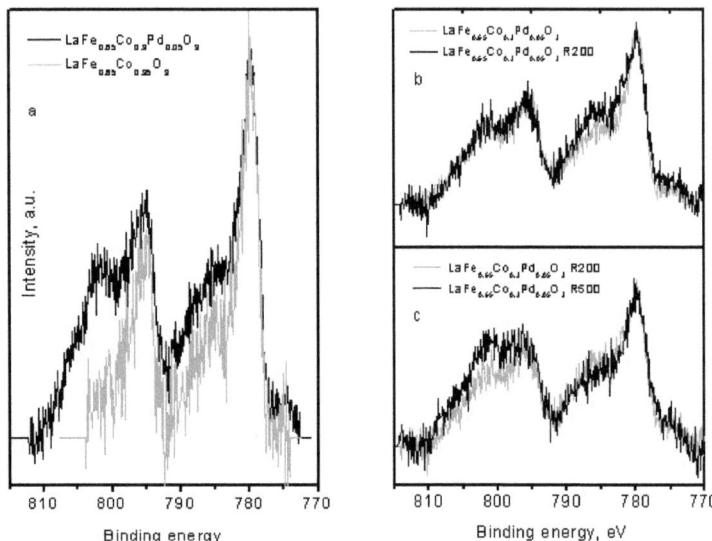

Figure 5.28 XPS lines of cobalt in the Pd-free perovskite and Pd-substituted perovkite as prepared (a), after reduction at 200°C (b), and after reduction at 500°C (c).

The elemental ratio between Fe, Co, and La is almost identical to the stoichiometric ratio in both samples, i.e. the Pd-free perovskite LaFeCo$_{(0.35)}$ and Pd-impregnated perovskite Pd-LaFeCo$_{(0.35)}$. After reduction of the impregnated sample Pd-LaFeCo$_{(0.35)}$ at 200 °C, the Fe surface concentration seems to be somewhat increased. On the contrary, the ratio between Fe and Co changes in the sample with Pd-substituted perovskite after reduction (see Table 5.5). It is around 1:1 in the Pd-substituted perovskite whereas it is clearly larger than one in the Pd-free samples indicating a change in the lattice composition, being predominantly in the iron position. The Fe/La and Co/La ratios are surprisingly low after reduction at 500°C of the Pd-substituted perovskite LaFeCo$_{(0.3)}$-Pd indicating enrichment of lanthanum at the catalyst surface in the reduced and oxidized state.

Palladium states in the lanthanum based perovskites

The palladium lines are clearly asymmetric (Fig. 5.29), in particular in those samples containing palladium in the crystal lattice i.e. Pd-substituted perovskites (Fig. 5.30). As line asymmetry is a feature confined to the metallic state, this asymmetry indicates the presence of additional ionic palladium states. There are at least three ionic Pd states present in both types of perovskites.

State I has a BE between and 336.5 eV, it is present in all initial samples and disappears after reduction treatments. Pd^{2+} is the most plausible explanation for this state. The BE for Pd^{2+} is reported to be equal to 336.5 eV [74].

State II with a BE between 337.3 and 337.8 eV was found in all initial samples. State II disappears upon reduction already at 200°C for both perovskites $LaFeCo_{(0.3)}$-Pd and Pd-$LaFeCo_{(0.35)}$. Pd $3d_{5/2}$ BE of approx. 337 eV is often associated with palladium intra-crystalline [63, 71, 74].

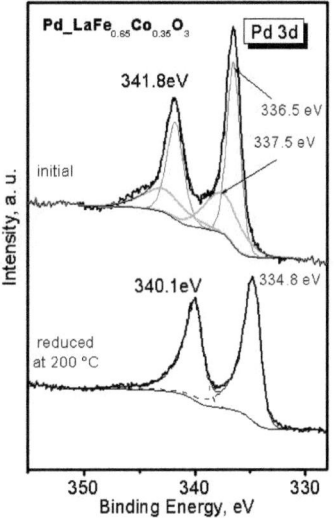

Figure 5.29 XPS lines of palladium in the Pd-supported perovskite Pd-LaFeCo$_{(0.35)}$ calcined in air at 700°C in the as prepared state (initial) and after reduction treatment at 200°C/2h.

State III has a BE of around 341 eV and may be due to palladium in oxidation state between $Pd°$ and Pd^{2+}. It is obvious from the spectra (see Fig. 5.30) that there is a doublet with exceptionally high binding energy. It is not present in the initial Pd-impregnated catalysts, and survives (at least partly) the reduction treatment at 500°C. Important to remember is that the BE may be influenced by Madelung field of the trivalent lanthanum and iron ions. State II would then arise from lattice palladium which is exposed at the surface where half of the Madelung stabilisation is missing ("surface lattice Pd ions"). Indeed, if the intra-crystalline fields are large enough to shift the BE by 4-5 eV, they may well cause a 1 eV shift in surface positions relative to PdO. The BE of palladium metal is about 335 eV and this value is in agreement with the BE for Pd $3d_{5/2}$ measured in the present work.

Table 5.5 Spectral parameters of elemental lines and surface concentration data: $LaFe_xCo_yO_3$ system.

Sample	La $3d_{5/2}$	Fe $2p_{3/2}$	Co 2p BE $(2p_{3/2})$	Co 2p ΔBE (1/2-3/2)	Co 2p (Sat/main)$_{3/2}$	O $1s^a$ BE	O $1s^a$ % of O 1s	Pd $3d$ $3d_{5/2}{}^b$	Pd 3d % of Pd 3d	Pd : Fe : Co : La
$LaFeCo_{(0.35)}$ R.T.	833.7	709.9	779.8	15.3	1.1	528.9 531.2	66 33	-	-	0 : 0.5_5 : 0.3 : 1
$Pd-LaFeCo_{(0.35)}$ R.T.	833.9	710.0	779.7	15.9	1.4	529.1 531.0	25 75	I : 336.5//1.4 II : 337.5//3.9	60 40	2.9 : 0.7 : 0.4 : 1
$Pd-LaFeCo_{(0.35)}$ R200	834.1	710.1	n. d.	n. d.	n. d	529.0 531.5	37 63	$334.8//1.6^c$	100	1.7_5 : 0.9 : ? : 1
$LaFeCo_{(0.3)}$-Pd R. T.	833.6	709.8	779.8	15.6	2.5	528.7 531.1	60 40	I: 336.4//1.4 II: 337.3//3.4 III: 341.9//5.9	25 40 35	0.1_5 : 0.5_5 : 0.5 : 1
$LaFeCo_{(0.3)}$-Pd R 200	833.6	709.8	779.9	15.9	4.4	528.8 531.2 525.8 (?)	66 34 -	I: $335.0//2.3^c$ - III: 341.0//4.0	55 - 45	0.09 : 0.4_5 : 0.5 : 1
$LaFeCo_{(0.3)}$-Pd R 500	834.0	709.9	779.8	15.7	2.6	528.7 531.1 526.0 (?)	70 30 -	I: $335.2//2.0^c$ III: 340.8//4.8	50 50	0.08 : 0.3_5 : 0.3 : 1

[a] probably influenced by contribution from C-O species
[b] BE // FWHM
[c] Asymmetry parameters
R. T.: Room Temperature
n. d.: not detected
(?) Weak signal

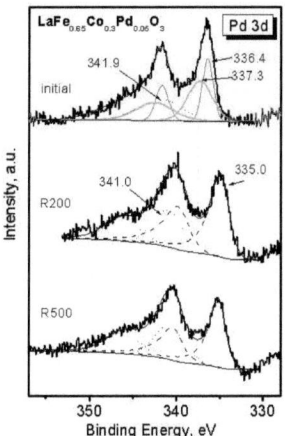

Figure 5.30 XPS lines of palladium in the Pd-substituted perovskite LaFeCo$_{(0.3)}$-Pd, calcined in air at 700°C/3h as-prepared (initial) and after reduction treatments in 4.2 vol. % H$_2$ + Ar at 200°C (R200) and at 500°C/2h (R500).

Redox behaviour of the perovskite BaTi-Pd by means of XPS

Fig. 5.31 displays the XPS lines of the perovskite BaTi-Pd-700°C. The Ba 3d lines for all the perovskites are all around 779 eV (Fig. 5.31 b). This result is in good agreement with BEs for Ba 3d$_{5/2}$ reported in the open literature [93]. Ti 2p$_{3/2}$ lines (Fig.5.31 a) are all around 457.8 eV with exception of the perovskite BaTi-Pd after reduction treatment at 500°C. Similar values for Ti 2p$_{3/2}$ lines are found in the literature as well [93, 94]. Ba and Ti XPS lines are not really affected with reduction treatment up to 500°C, thus showing the crystal stability of the perovskite even under the given reduction conditions.

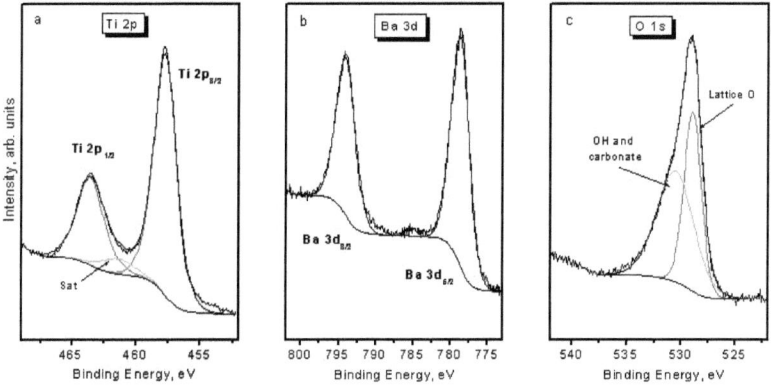

Figure 5.31 XPS lines of the perovskite BaTi-Pd calcined in air at 700°C/3h, as-prepared.

There are two oxidation states for oxygen (Fig. 5.31 c). The first state with BE = 529 eV belongs to oxygen from the crystal lattice. The second oxidation state of oxygen O 1s with BE = 530 eV is related to the presence of carbonates. This observation is supported by the XRD results reported in section 5.1.2. Low concentration of $BaCO_3$ was detected in the BaTi-Pd catalyst after calcination in air at 700°C/3h. The BE is small and therefore reflects the strong basicity of the Ba-containing surface.

The Ba/Ti ratio remained almost constant in both samples, i.e. the Pd-free perovskite BaTi calcined in air at 700°C and in the palladium impregnated perovskite Pd-BaTi after heat-treatment in air at 700°C. The presence of palladium did not cause significant modifications in the surface concentration of the other elements. The perovskite with palladium in the crystal lattice BaTi-Pd-700°C showed a similar Ba/Ti ratio compared to the Pd-free perovskite calcined in air at 700°C BaTi. Barium enrichment at the surface was observed for the perovskite Pd-BaTi-700°C after reduction treatment at 200° and 500°C. Under similar reduction conditions the same occurs to BaTi-Pd although to a much smaller extent (Table 5.6).

Palladium state in the barium based perovskites

Already in their initial stages, the Pd lines of the BaTi based samples are clearly asymmetric (Figs. 5.32), particularly those samples containing palladium in the lattice (Fig. 5.33). Line asymmetry is a feature that indicates the presence of additional ionic palladium states. At least three ionic palladium states were found in the Pd-free, and Pd-substituted perovskites BaTi-Pd.

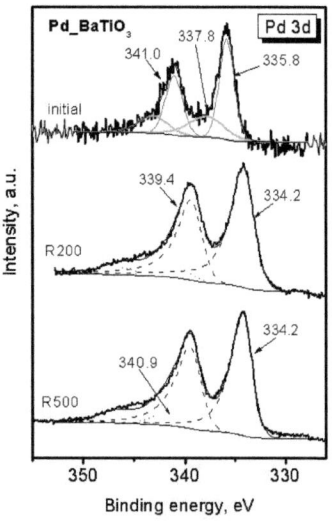

Figure 5.32 XPS lines of palladium in the supported perovskite Pd-BaTi calcined in air at 700°C/3h (initial), and after reduction treatment in 4.2 vol. % H_2 + Ar at 200°C(R200) and at 500°C/2h (R500).

Table 5.6 Spectral parameters XPS lines and surface concentration data of BaTi-Pd system*.

Sample	Ba $3d_{5/2}$	Ti $2p_{3/2}$	O 1s [a] BE	% of O 1s	Pd 3d $3d_{5/2}$ [b] 3d	% of Pd	Pd : Ba : Ti
BaTi R. T.	778.8	457.8	529.0 530.9	40 60	-	-	0 : 0.5_5 : 1
Pd-BaTi R. T.	778.8	457.8	528.9 531.5	50 50	335.8//1.7 337.8//4.3	67 33	0.2_5 : 0.7 : 1
Pd-BaTi R200	778.7	457.8	529.3 532.5	90 10 (?)	334.2/1.6 [c] 341.0//4.0	89 11	0.1_5 : 1.0 : 1
Pd-BaTi R500	779.2	457.9	529.2 530.6	25 75	334.3//2.2 [c] 340.9//4.1	81 19	0.2 : 1.6 : 1
BaTi-Pd R. T.	778.8	457.8	529.0 530.6	45 55	336.0//2.8 340.9//4.5	67 33	0.07_5 : 0.5_5 : 1
BaTi-Pd R 500	778.4	457.2	529.2 530.6	25 75	334.7//3.3c [c]	100	0.05 : 0.7 : 1

* All powder samples calcined in air at 700°C/3h
[a] probably influenced by contribution from C-O species
[b] BE // FWHM
[c] Asymmetry parameters
R. T.: Room Temperature

Similar to the perovskite LaFeCo$_{(0.3)}$-Pd, the catalyst BaTi-Pd showed three palladium oxidation states as mentioned above. State I with BE of 335.8, it is present in all as- synthesized supported perovskites (Pd-BaTi) but missing after reduction (Fig. 5.32). These binding energies are associated with the presence of Pd^{2+}. The Pd 3d$_{3/2}$ BEs of 336.5 eV for PdO is reported in the literature [74]. The palladium oxidation state may be modified depending on the host perovskite.

State II with BEs between 337.3 and 337.8 eV is associated with palladium intra-crystalline [74]. It was observed in all initial samples except for BaTi-Pd (Fig. 5.33). This does not necessarily exclude its presence in small quantities due to the presence of a third component. The relatively wide XPS peaks of BEs 336 and 340.9 eV may overlap any other signal between them (see Fig. 5.33). The wide XPS signals observed in the spectrum suggest the presence of other palladium species.

State III of BE around 341 eV associated with a modified palladium oxidation state between Pd° and Pd^{2+}. It is quite obvious from the spectra that there is a doublet with exceptionally high binding energy (> 343.5 eV). BEs at about 343.5 eV are associated with Pd^{4+} [71, 74]. BEs may be influenced by Madelung field of the tetravalent Ti-ions.

The BE of palladium metal is reported in the literature at around 335 eV [71, 74]. A Pd° BE of 334.7 eV was observed in BaTi-Pd after reduction treatment at 500°C/2h. Abnormal binding energies of about 334.3 eV (Pd°) were measured after reduction of Pd-BaTi (Table 5.6).

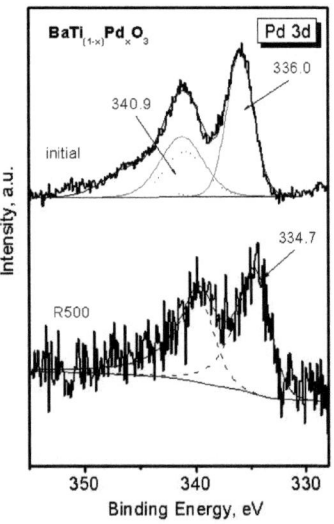

Figure 5.33 XPS lines of palladium in the perovskite BaTi-Pd calcined in air at 700°C/3h (initial), and after reduction treatment in 4.2 vol. % H_2 + Ar at 500°C (R500).

5.2 Catalytic coating development

EB-PVD PYSZ-coating

Standard partially stabilized zirconium oxide (PYSZ) coatings were first deposited on alumina plates from the 7 wt. % Y_2O_3 stabilized ZrO_2 ingots by means of EB-PVD method. Fig. 5.34 shows the typical columnar microstructure of the as-coated EB-PVD PYSZ-coatings at the top surface (a) and at the cross-section of the coating (b). The microstructure reveals a columnar morphology with a column diameter of about 20 µm. The thickness of the PYSZ ceramic substrates was around 200 µm.

(a) (b)

Figure 5.34 FE-SEM images of the EB-PVD PYSZ coating as coated, top surface (a) and cross section (b).

Table 5.7 Experimental conditions for EB-PVD PYSZ coated with the $LaFeCo_{(0.3)}$-Pd active phase.

Coating method No.	Experimental approach	Sample code
Direct Sol-coating method	-Partially hydrolyzed gel with a perovskite composition of $LaFeCo_{(0.3)}$-Pd	PYSZLaFe-1
Particulate coating method	-Alcohol-based suspension mixed with powder $LaFeCo_{(0.3)}$-Pd, which was previously calcined at 500°C for 3 h and milled during 60 min	PYSZLaFe-2
Particulate + Sol-coating method	-Aqueous suspension loaded with powder $LaFeCo_{(0.3)}$-Pd, which was previously calcined at 500°C for 3 h and milled during 60 min + partially hydrolyzed gel with a composition of $LaFeCo_{(0.3)}$-Pd	PYSZLaFe-3

The EB-PVD PYSZ substrates were coated with a selected perovskite composition by employing three different experimental approaches (see Table 5.7). A partially hydrolyzed gel, which was prepared via the same citrate route reported in section 4.1.1. was employed to integrate the perovskite on the EB-PVD PYSZ substrates. The particle size of the powder particles $LaFeCo_{(0.3)}$-Pd was reduced by milling to less than 10 µm. According to the gained practical experience INTERKAT® recommended for catalyst to be deposited on ceramic substrates a particle size of 10 µm and below. Fig. 5.35 shows the particle size and morphology of powder $LaFeCo_{(0.3)}$-

Pd-500°C milled with ZrO_2 balls in 2-propanol, in the relation of 1:5:10 (catalyst: alcohol: ZrO_2 balls) within the intervals of 20, 40 and 60 minutes. The particle size was strongly decreased within the first 20 minutes of milling but particles with relatively large sizes were still present (>10 µm). After 40 minutes of milling most of the particle sizes were below 10 µm. A uniform particle distribution was obtained after 60 min of milling under the same conditions image (c) Fig. 5.35. Particle size determination proved that 87 % of the particles sizes were below 5 µm and approx. 65 % were below 1 µm.

(a) (b) (c)

Figure 5.35 Particle morphologies and distribution of the powder $LaFeCo_{(0.3)}$-Pd calcined at 500°C for 3 h and milled in 2-isopropanol for 20 min (a), 40 min (b) and 60 min (c).

5.2.1 Microstructure and phase analysis of the perovskite after the impregnation and calcination of the PYSZ coating

The first approach to synthesize and integrate the EB-PVD PYSZ substrate with the perovskitic catalyst was carried out by direct impregnation of the ceramic EB-PVD coating with the perovskite gel previously synthesized via the citrate route as reported in the experimental section (Sample code PYSZLaFe-1). The EB-PVD PYSZ coatings in the as-prepared state were immersed into the partially hydrolyzed gel, kept inside for a few minutes and then taken out, followed by elimination of surface excess solution. Samples were dried in two steps to slowly evaporate the solvent; in the first step the coatings were dried at atmospheric air for a few minutes and then, in a second step in oven at 80°C for at least 3 h. The procedure was repeated three times, and the so-obtained coatings were calcined under static air at 900°C for 3 h. It is observed that lower calcination temperatures resulted in very poor adhesion between gel and PYSZ ceramic substrates. Fig. 5.36 displays top coat SEM images of the perovskite deposited EB-PVD PYSZ coating calcined in air at 900°C for 3 h. The XRD pattern of the perovskite deposited EB-PVD PYSZ coating shows the stability of the perovskite structure after calcination in static air at 900°C for 3 h (see Fig. 5.37).

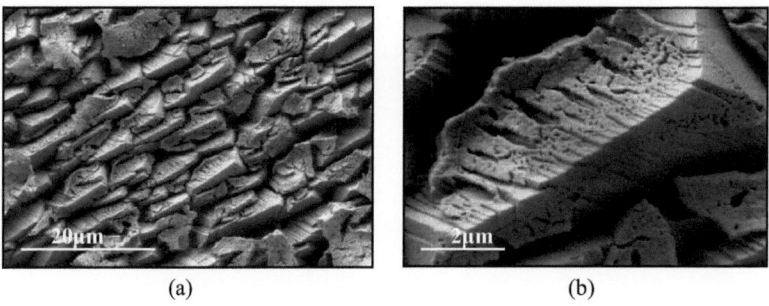

Figure 5.36 Top-coat images of the PYSZ EB-PVD-perovskite integrated coating calcined at 900°C/3h, 2000X (a) and 13000X (b), (Sample code PYSZLaFe-1).

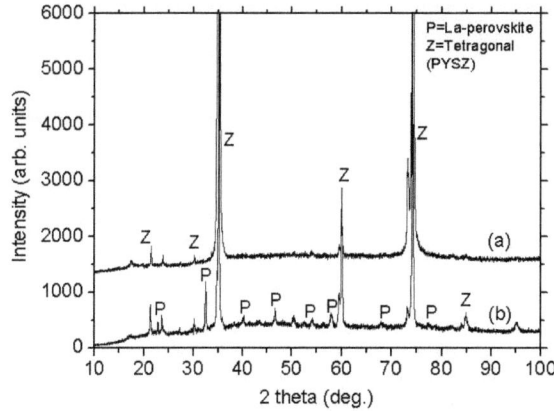

Figure 5.37 XRD patterns of the EB-PVD PYSZ coatings as prepared (a), and after coating with the La-based perovskite and calcined at 900°C/3h (b).

Alternatively, the EB-PVD PYSZ substrate was coated with a perovskite catalyst in vacuum in order to achieve a better coverage of the EB-PVD PYSZ columnar structure and to fill the pores between the columns. This method requires suspensions with extremely low solid concentrations and consequently low solution viscosities, so that the gel can flow into the pores and between the columns of the EB-PVD PYSZ ceramic coatings. Due to the low concentration of the suspension employed to coat the ceramic EB-PVD PYSZ coatings, the method applying vacuum has to be repeated several times. By repeating this step seven times, the amount of catalyst deposited on the surface of the ceramic coating was increased. Even though the multiple coating procedures employed the amount of the final deposited catalyst was very small (e.g. 1 mg catalyst/g PYSZ coating).

The second approach to deposit La-based perovskite catalyst on EB-PVD PYSZ coatings was the direct impregnation by employing an aqueous suspension (with 10 wt. % of solid loading) of the previously synthesized perovskite particles, which were calcined in air at 500°C for 3 h

(Sample code PYSZLaFe-2). Fig. 5.38 shows the top coat images and cross section of the perovskite catalyst deposited EB-PVD PYSZ coating prepared by this method. The amount of catalyst achieved with this experimental method was less than 2 mg perovskite-catalyst/g PYSZ coating, and the La-perovskite particles were not really fixed to the EB-PVD PYSZ coating surface. The perovskite particles were attached to the ceramic substrate only by adhesion forces. This method provides an interesting approach to uniformly coat the PYSZ columns.

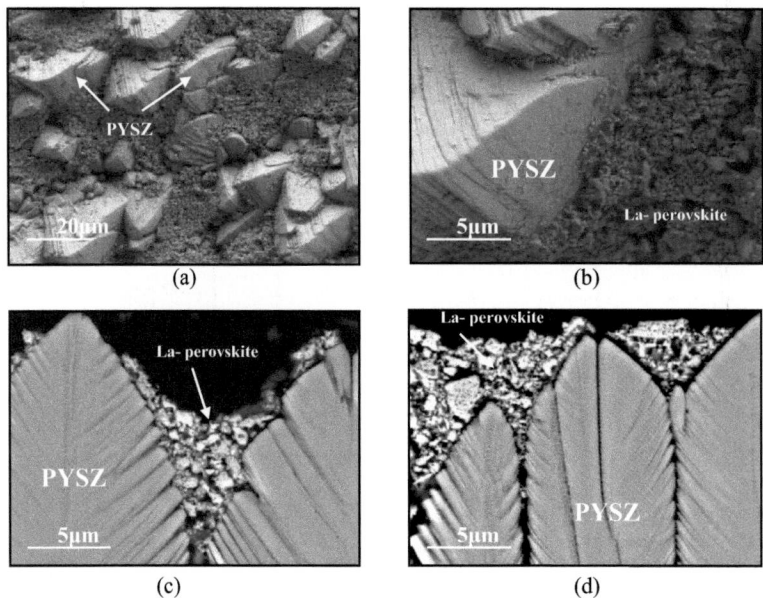

Figure 5.38 Top coat (a) and (b), and cross section (c) and (d) of the PYSZ EB-PVD-perovskite coating (900°C/3h air calcined) prepared via the impregnation method from a solid loaded aqueous suspension. Solid particles: La-based perovskite previously calcined at 500°C/3h, (Sample code PYSZLaFe-2).

The third method used to deposit La-based perovskite catalyst on to the EB-PVD PYSZ coating was the direct impregnation of the EB-PVD PYSZ coating with a suspension based on a mixture of the perovskite sol containing the La-perovskite particles which were pre-synthesized by sol-gel, calcined at 500°C/3h and milled for 60 min as described in this section (Sample code PYSZLaFe-3). Final coatings were loaded with approx. 2.8 mg catalyst/g PYSZ coating after calcination in air at 900°C/3h. This method yields a more homogeneous distribution and higher coating degree of catalyst loading in comparison with the first two previously mentioned coating methods (SEM-images (a) and (b) in Fig. 5.39). The use of partially hydrolyzed gel in the third method acts as a "glue" creating a better contact effect between particles-particles and particles-substrate. Some areas of the PYSZ columns were coated with a thin La-perovskite layer with a thickness below 200 nm (see image (d) Fig. 5.39). The XRD pattern of the PYSZ in the as coated state displays the preferred crystallographic growth in the <110> directions (see Fig. 5.37 a). The morphology and orientation of the columns depends primarily on the relative position of the substrates to the vaporisation source, i.e. vapour incidence angle (VIA), which is the angle between the substrate normal and the average direction of vapour incidence [95]. Peaks corresponding to the presence of the orthorhombic crystal structure are observed at the PYSZ coatings loaded with the La-based perovskite (see Fig. 5.37 b) and no other oxide species were detected.

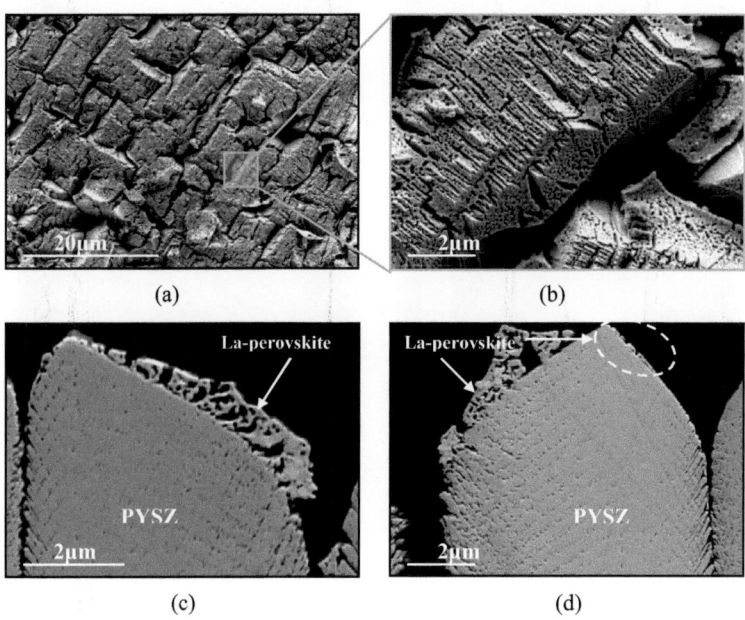

Figure 5.39 Top coat (a) and (b) and cross sections (c) and (d) of the PYSZ EB-PVD-perovskite coating calcined in air at 900°C/3h, prepared via direct impregnation with a suspension of partially hydrolyzed La-based gel + La-based particles, (Sample code PYSZLaFe-3).

Cyclic thermal ageing of the system coatings $LaFeCo_{(0.3)}$-Pd-EB-PVD PYSZ

The coatings $LaFeCo_{(0.3)}$-Pd-EB-PVD PYSZ were thermally aged under static air at 900°C for 30 min and then were cooled down to room temperature in 10 min. The purpose of these experiments was to prove the thermal shock resistance of the catalytic coatings under extreme conditions. The same thermal cycle was applied to the coatings $LaFeCo_{(0.3)}$-Pd-EB-PVD PYSZ at 500°C. Following a 30 min dwelling at 500°C the samples were cooled down to room temperature in 10 min. SEM-micrographs in Figs. 5.40 to 5.42 show the top surface of the coatings $LaFeCo_{(0.3)}$-Pd-EB-PVD PYSZ after 25, 50 and 100 ageing cycles at 500°C. No weight loss was observed during the thermal ageing of the coatings at this temperature. The XRD analysis of the coatings $LaFeCo_{(0.3)}$-Pd-EB-PVD PYSZ after ageing at 500°C showed no phase(s) transformation in both the perovskite and the coatings EB-PVD PYSZ (see Fig. 5.43).

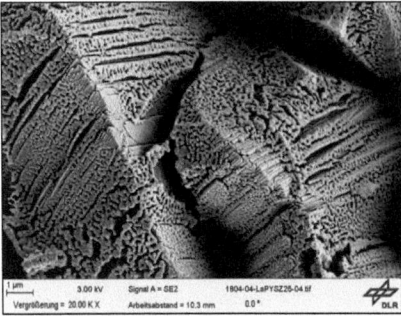 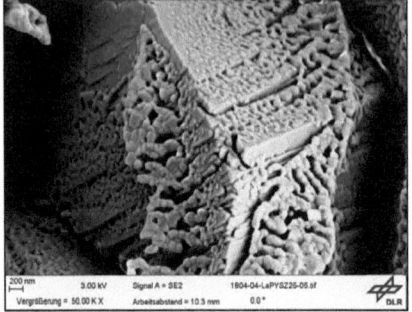

Figure 5.40 Thermal ageing in air at 500°C (25 cycles) of the coating LaFeCo$_{(0.3)}$-Pd-EB PVD PYSZ, 20000 X (left) and 50000 X (right).

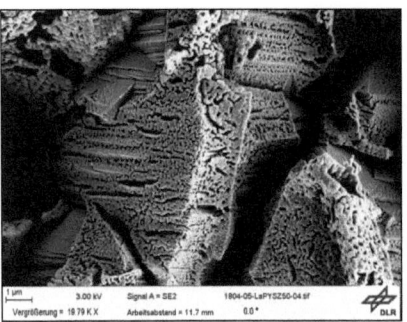 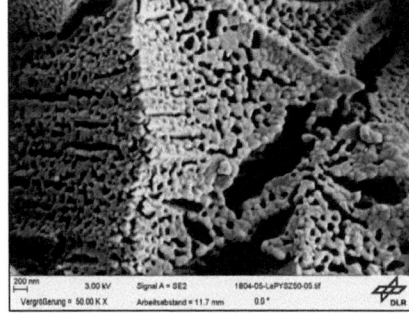

Figure 5.41 Thermal ageing in air at 500°C (50 cycles) of the coating LaFeCo$_{(0.3)}$-Pd-EB PVD PYSZ, 20000 X (left) and 50000 X (right).

 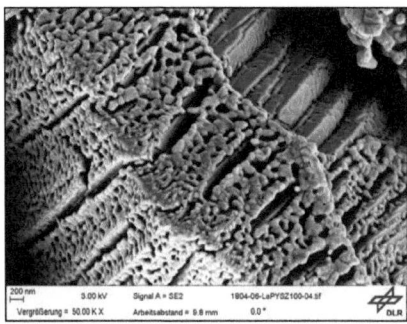

Figure 5.42 Thermal ageing in air at 500°C (100 cycles) of the coating LaFeCo$_{(0.3)}$-Pd-EB PVD PYSZ, 20000 X (left) and 50000 X (right).

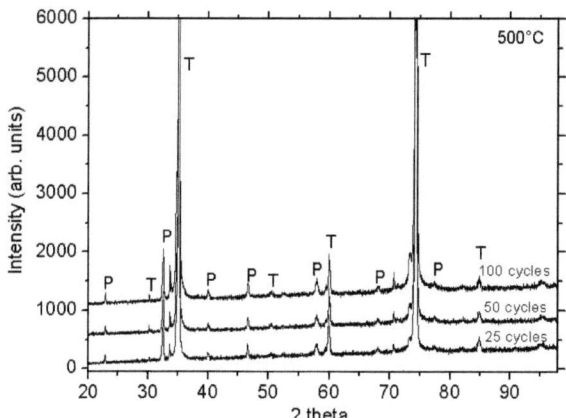

Figure 5.43 XRD patterns of the coatings $LaFeCo_{(0.3)}$-Pd-EB PVD PYSZ after thermal ageing under static air at 500°C, P: Perovskite and T: Tetragonal PYSZ.

Thermal ageing at 900°C caused some changes in the microstructure of the catalytic coating (see Figs. 5.44-5.46). Particularly on top of the PYSZ columns, it is clearly seen that the small perovskite grains start to migrate downwards at this temperature. Sintering begins with the consolidation of the bonding necks between perovskite grains that tend to migrate to the column bottom. This phenomenon was not observed at the coatings $LaFeCo_{(0.3)}$-Pd-EB PVD PYSZ thermally aged at 500°C. No weight loss of the catalytic coatings after cyclic thermal ageing was observed. This indicates that the perovskite particles remain on the EB-PVD PYSZ ceramic surface, but sinter together. In order to overcome the thickness reduction and elimination by the migration of perovskite particles, a thermal treatment can be carried out between the impregnation steps. This also indicates that the coatings $LaFeCo_{(0.3)}$-Pd were bonded well to the EB-PVD PYSZ and displayed good thermal shock resistance.

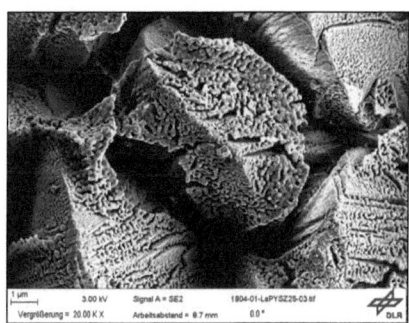 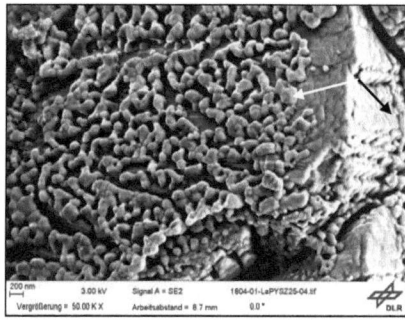

Figure 5.44 Thermal ageing in air at 900°C (25 cycles) of the coating $LaFeCo_{(0.3)}$-Pd-EB PVD PYSZ, 20000 X (left) and 50000 X (right).

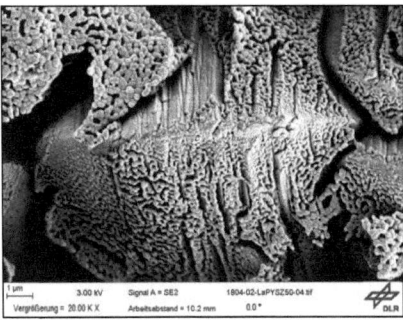 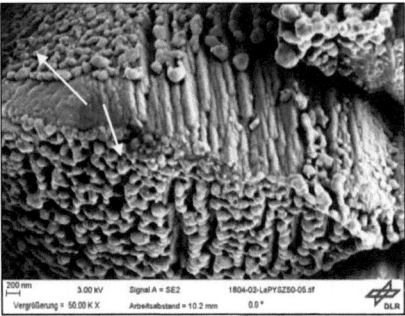

Figure 5.45 Thermal ageing in air at 900°C (50 cycles) of the coating LaFeCo$_{(0.3)}$-Pd-EB PVD PYSZ, 20000 X (left) and 50000 X (right).

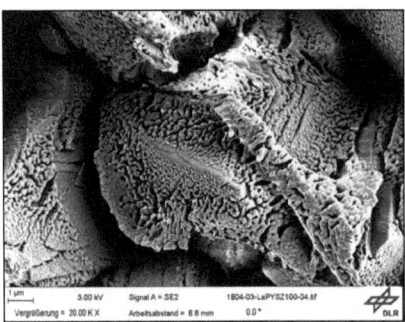 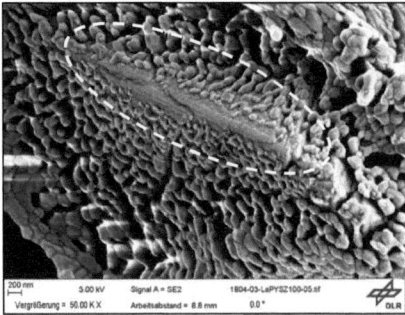

Figure 5.46 Thermal ageing in air at 900°C (100 cycles) of the coating LaFeCo$_{(0.3)}$-Pd-EB PVD PYSZ, 20000 X (left) and 50000 X (right).

Figure 5.47 shows the results obtained by XRD measurements of the coatings LaFeCo$_{(0.3)}$-Pd-EB PVD PYSZ after cyclic thermal ageing at 900°C. The spectra display only minimal changes in the intensity signals that correspond to the lanthanum based perovskite. The perovskite structure was not modified after 25, 50 and 100 cycles of thermal ageing at 900°C. The EB-PVD PYSZ coatings show a preferred orientation in the <110> direction. No crystal phase(s) changes were observed after thermal ageing at 900°C.

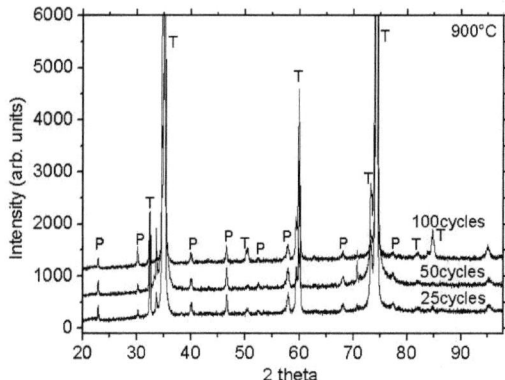

Figure 5.47 XRD patterns of the coatings LaFeCo$_{(0.3)}$-Pd-EB PVD PYSZ after thermal ageing under static air at 900°C, P: Perovskite and T: Tetragonal PYSZ.

5.2.2 Perovskite catalyst coated on top of monolithic cordierite substrate

Cordierite substrates with a volume of about 36.6 ml and average weight of 15.82 g provided by INTERKAT© were coated with the La-based perovskite catalyst via the direct impregnation method. An aqueous suspension was prepared with 40 wt. % of solids based on La-based perovskite particles, previously calcined at 500°C/3h and milled for 60 min, were mixed with colloidal solution of SiO_2. Coating of the cordierite substrate was carried out by casting of the perovskite based suspension. After coating process the excess La-perovskite based solution was eliminated with pressurized air. The same procedure was repeated three times and the so-obtained cordierite supported catalyst was then dried by heating in hot air at 150°C until no weight-loss was observed.

Figure 5.48 Cordierite support coated with La-based perovskite + colloidal silica suspension calcined in air at 700°C/3h (a) and dimensions (b).

Fig. 5.48 displays the image of the cordierite substrate coated with the La-perovskite catalyst and its dimensions. Approx. 1.45 % weight-loss of the coated substrate was registered on calcination in air at 700°C for 3 h. Final amount of La-perovskite catalyst fixed on the substrate surface was approx. 3.72 g. The wall- thickness of the cordierite support was around 200 μm and the La-perovskite coating thickness approx. 10 - 20 μm [see Figs. 5.49 (a) and (b)]. The dimensions of the cordierite substrate, wall-thickness and channel aperture (approx. 1.2 mm together) provide a ceramic monolith with 400 cpsi (cpsi = cells per square inch).

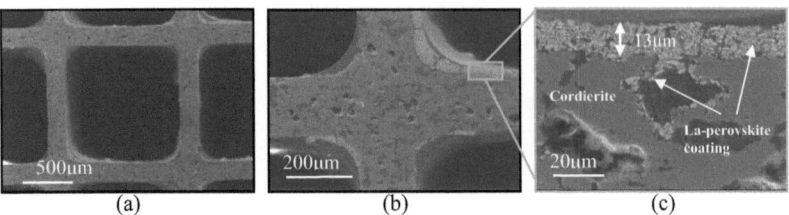

Figure 5.49 SEM-images of cordierite support cross-sections (a) and (b), and La-perovskite coating thickness supported on cordierite wash-coat (c).

Microstructure characteristics of the coated La-perovskite catalyst on the top-surface are shown in Fig. 5.50. A highly homogenous perovskite based coating was obtained. Small islands of SiO_2 rich zones were also observed [SEM-image (a) in Fig. 5.50]. The use of SiO_2 colloidal solution functions well as "glue" providing good contact and adhesion in the particle-particle and particle-substrate interactions. The zones building crack networks as observed at low magnifications (700 X) were filled with the SiO_2 particles as the high magnifications reveal (see images (b) and (c) Fig. 5.51). The fracture surface confirms the homogeneity of the catalytic coating on the substrate surface. The La-based perovskite coating can be easily differentiated from the substrate surface (Fig. 5.51). This Figure shows the fracture surface of one wall of the cordierite substrate and the detail inside one coated channel surface.

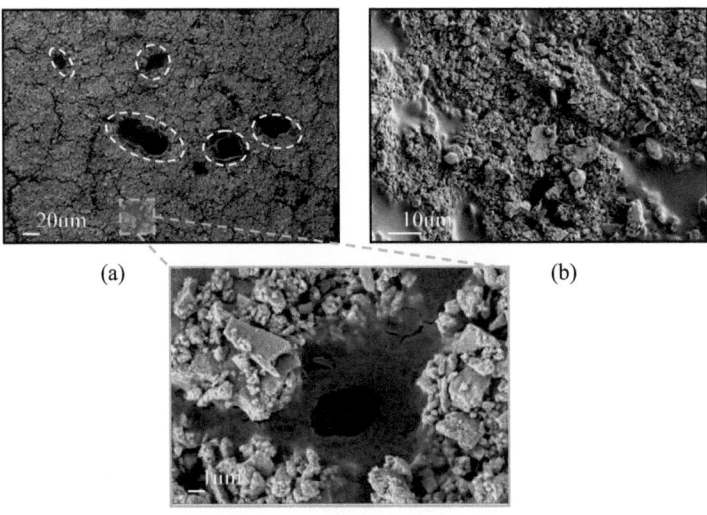

Fig 5.50 SEM images of the catalytic coating surface with composition $LaFeCo_{(0.3)}$-Pd on the cordierite substrate after calcination in air at 700°C/3h.

Figure 5.51 Fracture surface of the cordierite substrate coated with La-perovskite catalyst.

5.3 NO_x-reduction capability and N_2-selectivity of the powder catalysts

The catalyst formulations tested for NO_x-reduction under lean conditions using H_2 as reducing agent are reported in Table 5.8. The catalysts were tested under the same reaction conditions, feed composition and temperature program. Similarly, the effect of CO and H_2O + CO_2 in the gas stream was analyzed. In order to prove the catalytic performance of the Pd-doped perovskites, several catalyst compositions were tested under 0.072 vol. % NO / 5 vol. % O_2 / 1 vol. % H_2 gas mixtures. Basically perovskite catalysts with substitution of two atoms at the A- or B- site (double site function catalyst) and perovskite catalyst with one atom at each A- and B-site (single site function catalyst) were tested. The BET specific surface areas of the as-synthesized catalysts are reported in Table 5.8. Detailed description of the catalysts synthesis is given in the experimental section of chapter 4. Table 5.9 summarises the number of catalysts tested in this work and the reaction conditions of each test using H_2 as reducing agent. The NO and NO_2 signals were continuously monitored using the chemiluminescent technique NO/NO_x. The feed stream was continuously analyzed during the NO_x-reduction reaction with FT-IR spectroscopy. Since no NH_3 was formed under lean conditions, the N_2 concentration was calculated via the difference between feedstock and effluent concentrations of the N-containing species.

Table 5.8 BET-specific surface areas and phase composition (XRD) of the perovskite based catalysts calcined at 900°C/3h and tested at CenTACat.

Catalyst [c]	SSA [d] ($m^2.g^{-1}$)	Phase (s)
$LaFeCo_{(0.475)}$-Pd	2.113	$LaFe_{0.6}Co_{0.4}O_3$ perovskite+Co_3O_4[a]+PdO[b]
$LaFeCo_{(0.475)}$-Rh	-----	$LaFe_{0.6}Co_{0.4}O_3$ perovskite+Co_3O_4[a]
$LaFeCo_{(0.3)}$-Pd	1.657	$LaFe_{0.6}Co_{0.4}O_3$ perovskite+PdO[b]
$LaCe_{(0.05)}FeCo_{(0.3)}$-Pd	-----	$LaFe_{0.6}Co_{0.4}O_3$ perovskite+ CeO_2[b]+Fe_2O_3[a]
$LaCe_{(0.05)}Fe$-Pd	1.598	$LaFeO_3$ perovskite+CeO_2[a]+Fe_2O_3
$LaCe_{(0.4)}Fe$-Pd	6.647	$LaFeO_3$ perovskite+CeO_2+Fe_2O_3
BaTi-Pd	21.33	$BaTiO_3$ perovskite+$BaCO_3$[b]

[a] traces
[b] small amount
[c] the formula implies the theoretical catalyst composition
[d] SSA: Specific Surface Area

Table 5.9 Perovskite compositions tested during H_2-SCR of NO_x at CenTACat.

Reaction mixture → Catalyst	$NO/O_2/H_2/He$ [a]	$NO/O_2/H_2/CO_2/H_2O/He$ [b]	$NO/O_2/H_2/CO/He$ [c]
LaFeCo$_{(0.475)}$-Pd	τ	τ	τ
LaFeCo$_{(0.475)}$-Rh	τ	-----	-----
LaFeCo$_{(0.3)}$-Pd	τ	τ	τ
LaCe$_{(0.05)}$FeCo$_{(0.3)}$-Pd	τ	τ	τ
LaCe$_{(0.05)}$Fe-Pd	τ	τ	τ
LaCe$_{(0.4)}$Fe-Pd	-----	τ	-----
BaTi-Pd	τ	τ	τ

τ = tested under the selected reaction conditions
[a] 720 ppm NO, 5 % O_2, 1 % H_2, He balance
[b] 720 ppm NO, 5 % O_2, 1 % H_2, 7.2 % CO_2, 7.2 % H_2O, He balance
[c] 720 ppm NO, 5 % O_2, 1 % H_2, 0.25 % CO, He balance
W/F (weight to flow ratio) = 0.065 $g_{cat}.s.ml^{-1}$

5.3.1 NO_x-reduction with H_2 under excess oxygen

In Pd-free perovskites almost no NO_x-conversion was observed implying poor participation of the perovskitic bulk structure in the reaction mechanism. This result confirms that the NO conversion occurs through the palladium surface alone. Fig. 5.52 displays the catalytic activity results of the different La-based perovskites substituted with the transition elements (Fe, Co) and doped with palladium during the NO_x-SCR with H_2 reaction (0.072 vol. % NO + 5 vol. % O_2 + 1 vol. % H_2 + He). The catalysts start to reduce NO_x at 160°C with the maximum NO_x-conversion occurring between 180°C and 230°C. At 300°C and above the competing oxidation of hydrogen causes a decrease in the NO_x-conversion. Three competing reactions are implied during the NO_x-reduction with H_2 under lean conditions; the reduction of nitrogen oxide to nitrogen as giving with reaction (1), the simultaneous formation of dinitrogen oxide (N_2O) according to reaction (2), and the hydrogen combustion as giving with reaction (3).

$$2NO + 4H_2 + O_2 \rightarrow N_2 + 4H_2O \quad (1)$$

$$2NO + 3H_2 + O_2 \rightarrow N_2O + 3H_2O \quad (2)$$

$$O_2 + 2H_2 \rightarrow 2H_2O \quad (3)$$

Fig. 5.52 reveal the complex characteristics of the catalytic surface properties due to the presence of competitive and successive reactions leading to the formation of N_2, N_2O, H_2O and NH_3, ammonia being detected only for rich gas mixture (no oxygen in the feed). It is likely that three competing reactions occur during the NO_x-reduction with H_2. As Fig 5.2 displays, the NO_x conversion raises to a maximum of 74 % over LaFeCo$_{(0.3)}$-Pd at about 200°C, followed by a sharp decrease which levels off with 25 % at 300°C. Increased Co-content at the B- site of the catalysts (curve A) results in a higher NO_x-conversion (80 %) over a broader temperature range (175°-225°C).

The catalyst LaCe$_{(0.05)}$Fe-Pd has a NO_x-conversion maxima of 80 % at 200°-230°C (curve D in Fig. 5.52). It is possible that the partial substitution of the La-site with cerium tetravalent in the perovskite LaCe$_{(0.05)}$Fe-Pd increases the number of vacancies at the A-site of the perovskite due to the limited solid solution. On Co-addition in the catalyst LaCe$_{(0.05)}$FeCo$_{(0.3)}$-Pd, the maximum NO_x-conversion was slightly shifted to lower temperatures. However, curve c of Fig. 5.52 shows that the

NO_x-conversion decreases down to 56 % and reduces sharply down to 15 % from 115° to 130°C. Fig. 5.52 (curve C) shows a sharp increase in NO_x-conversion, which begins at 115°C, reaching to a maximum of 56 % at 180°C on the catalyst $LaCe_{(0.05)}FeCo_{(0.3)}$-Pd. On the other hand, BaTi-Pd (curve E in Fig. 5.52) displays a maximum NO_x-conversion of 70 % already at 170°C which is being the lowest NO_x-reduction temperature among the tested catalysts.

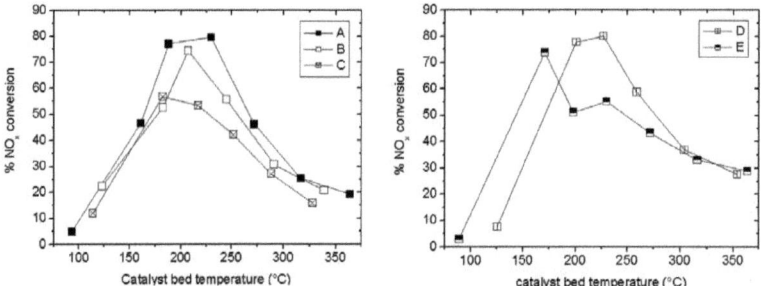

Figure 5.52 NO_x-reduction under lean conditions with; $LaFeCo_{(0.475)}$-Pd (A), $LaFeCo_{(0.3)}$-Pd (B), $LaCe_{(0.05)}FeCo_{(0.3)}$-Pd (C), $LaCe_{(0.05)}Fe$-Pd (D), BaTi-Pd (E), all catalysts calcined at 900°C/3h. Reaction conditions: 300 mg of catalyst, at a W/F = 0.065 $g_{cat}.s.ml^{-1}$, 720 ppm NO, 5 vol. % O_2, 1 vol. % H_2, He as balance.

One of the main issues when using H_2 as reducing agent is the unselective formation of relatively high amounts of N_2O, which is classified as a greenhouse emission and thus contributes to the global warming [96]. The concentration of N_2O being detected and calculated after numerical integration of the bands at 2202 and 2236 cm^{-1} from the IR-spectra taken after the NO_x-reduction reaction. Since no NH_3 was detected in the presence of excess oxygen, the N_2 concentration formed after the catalytic reduction was calculated via the difference between feedstock and effluent concentrations of the N-containing species. Fig. 5.53 displays the N_2-selectivity as a function of the catalyst bed temperature for a simple reaction mixture containing (0.072 vol. % NO + 5 vol. % O_2 + 1 vol. % H_2 + He). A complex mechanism involving competing reactions (i.e. the initial NO_x-reduction to N_2 and N_2O, and the hydrogen combustion) is responsible for the observed behaviour. Thus, the N_2-selectivities over the temperature range of 100° to 350°C a decline form with all the catalysts. The minimum and maximum N_2-selectivity varies with the catalyst composition. The decrease in cobalt content (Fe/Co ratio increase) of the catalysts had a positive effect on the N_2 selectivity at temperatures between 160°C and 205°C (see curves A and B Fig. 5.53). Although the NO_x conversion rate was not much affected by the decrease of cobalt, the selectivity of the catalyst was significantly modified. Above 200°C the selectivity to N_2 in the catalyst $LaFeCo_{(0.475)}$-Pd increases steadily to reach 80 % at 350°C. However, the lowest N_2-selectivity is detected with this catalyst at 200°C.

Maximum N_2–selectivity of the catalyst $LaCe_{(0.05)}FeCo_{(0.3)}$-Pd was 64.5 % at 182°C. N_2–selectivity decreased to a minimum of ~ 40.4 % at around 288°C. Further increase of the reaction temperature up to 328°C resulted in a slight increase in N_2-selectivity to 57.5 %. A very similar trend was observed with the Co-free catalyst $LaCe_{(0.05)}Fe$-Pd. Minimum N_2-selectivity with approx. 44 % was slightly shifted to lower temperatures (260°C) compared to the cobalt substituted catalyst (290°C).

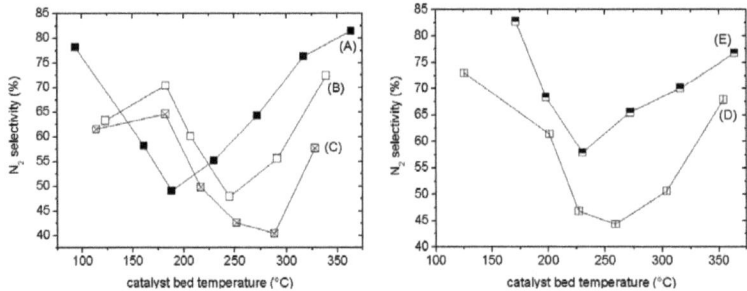

Figure 5.53 N_2-selectivity during the NO_x-reduction under lean conditions with; $LaFeCo_{(0.475)}$-Pd(A), $LaFeCo_{(0.3)}$-Pd (B), $LaCe_{(0.05)}FeCo_{(0.3)}$-Pd (C), $LaCe_{(0.05)}$Fe-Pd (D), BaTi-Pd (E), all calcined at 900°C/3h under static air. Reaction conditions: 300 mg of catalyst, at a W/F = 0.065 $g_{cat}.s.ml^{-1}$, 720 ppm NO, 5 vol. % O_2, 1vol % H_2, He as balance.

Maximum N_2-selectivity achieved with the BaTi-Pd-900°C based catalyst was 82.7 % at 170°C (curve E in Fig. 5.53). The selectivity to N_2 decreases at 230°C to a minimum of 57.9 % followed by an increase at 364°C to 76.8 %. In terms of N_2 production, the BaTi-Pd calcined at 900°C was the most selective catalyst between 177° and 230°C. It is likely that the crystal structure as well as the presence of barium facilitates the formation of nitrate complexes which further react with other N-species to form N_2. On the other side, if Pd is replaced with Rh an extremely poor NOx-conversion was observed with the Rh-substituted $LaFeCo_{(0.475)}$-Rh catalyst under the reaction conditions employed in this study (see Fig. 5.54). It is worthy to note that despite the lower catalytic activity N_2O was not produced with this catalyst.

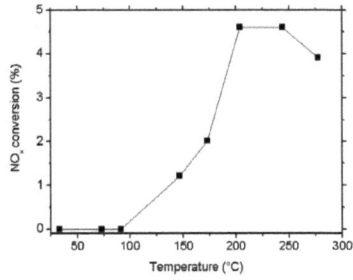

Figure 5.54 NO_x-reduction under lean conditions with $LaFeCo_{(0.475)}$-Rh calcined at 900°C/3h in static air. Reaction conditions: 300 mg of catalyst, at a W/F = 0.065 $g_{cat}.s.ml^{-1}$, 720 ppm NO, 5vol. % O_2, 1 vol. % H_2, He as balance.

5.3.2 NO_x-reduction with H_2 under lean conditions including H_2O and CO_2

The effect of H_2O_{vapour} and CO_2 on the NO_x-reduction under lean conditions is represented in Fig. 5.55. These data show that the NO_x-conversion with each catalyst is different and strongly affected with the addition of water vapour in the gas stream. A maximum NO_x-conversion of 57.3 % was achieved with the catalyst $LaFeCo_{(0.3)}$-Pd at approx. 220°C (curve B in Fig. 5.55). The same trend was observed for the catalyst with small cerium content $LaCe_{(0.05)}FeCo_{(0.3)}$-Pd (curve C in Fig. 5.55). The NO_x-conversion peak was slightly shifted to lower temperatures for this catalyst and a maximum NO_x conversion of 57.4 % was measured at 223°C. Fig. 5.55 displays an improvement in NO_x conversion due to the cobalt increase in the catalyst. 85 % of NO_x conversion was registered between 210°C and 245°C with the catalyst $LaFeCo_{(0.475)}$-Pd (curve A in Fig. 5.55).

The N_2-selectivity diagram on the right side of Fig. 5.55 shows the complex effect of the catalyst composition on the N_2-selectivity as a function of temperature. In terms of N_2 production, the catalyst $LaFeCo_{(0.3)}$-Pd was the most selective of these series under the given reaction conditions at temperatures above 250°C. The increase in cobalt content in the catalyst $LaFeCo_{(0.475)}$-Pd resulted in a slight decrease in N_2-selectivity (curve B in right side of Fig. 5.55). However, this catalyst showed a better performance during the NO_x-reduction in the presence of water vapour compared with the dry H_2-SCR reduction of NO_x. A small increase in selectivity to N_2 was observed over the catalyst $LaCe_{(0.05)}FeCo_{(0.3)}$-Pd between 195°C and 240°C. The N_2-selectivity of this catalyst is lower compared to cerium-free catalyst (curve A and C in right side graph in Fig. 5.55).

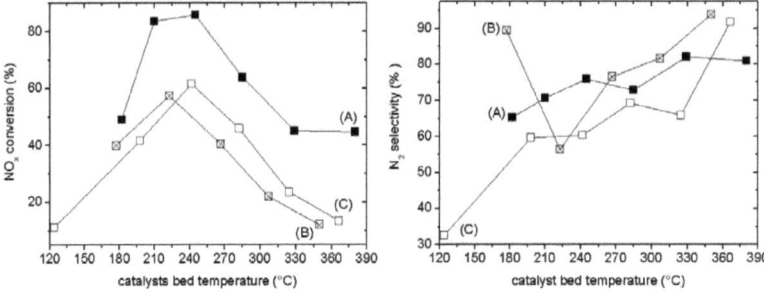

Figure 5.55 NO_x conversion (left) and N_2-selectivity (right) under lean conditions with; $LaFeCo_{(0.475)}$-Pd (A), $LaFeCo_{(0.3)}$-Pd (B), $LaCe_{(0.05)}FeCo_{(0.3)}$-Pd (C), all calcined at 900°/3h. Reaction conditions: 300 mg of catalyst, at a W/F = 0.065 $g_{cat}.s.ml^{-1}$, 720 ppm NO, 5 vol. % O_2, 1vol. % H_2, 7.2 vol. % H_2O, 7.2 vol. % CO_2, He as balance.

In this study, also a new composition base on BaTi-Pd was tested for NO_x-reduction under lean conditions in the presence of H_2O_{vapour} + CO_2. The NO_x-conversion results obtained with the BaTi-Pd catalyst are compared with the NO_x-conversions of the catalysts $LaCe_{(0.05)}$Fe-Pd and $LaCe_{(0.4)}$Fe-Pd in Fig. 5.56. Moreover, for comparison purposes, a platinum supported catalyst 1.0 wt. % Pt/SiO_2 provided by CenTACat was tested under the same reaction conditions. A wider NO_x conversion range lying between 160°C to 330°C was measured with the palladium substituted perovskite BaTi-Pd. In comparison to the dry NO_x-reduction, the catalyst showed around 15 % increase in the absolute NO_x conversion between 230°C and 270°C during the wet NO_x-reduction test (feed with H_2O_{vapour} + CO_2). A further increase in the catalyst test temperature resulted in a decrease at the NO_x-conversion. Right curve in Fig. 5.56 presents the high N_2-selectivity over the Pd-substituted catalyst during the NO_x-reduction in the presence of H_2O_{vapour} + CO_2. This catalyst displayed a slight improvement towards N_2 formation with water vapour in the gas stream (curve C in right graph of Fig. 5.56).

The catalytic performance of the BaTi-Pd catalyst (70 % NO_x conversion at 240°C) was slightly better compared to the catalyst $LaCe_{(0.05)}$Fe-Pd (curve A and C in Fig. 5.56). With the catalyst LaCe$_{(0.05)}$Fe-Pd the maximum NO$_x$-conversion (56 %) was reached at ~ 230°C. The reactivity of this catalyst was strongly affected by the presence of water vapour. The NO$_x$-conversion decreases by about 25% in the presence of water vapour in the feed over this catalyst.

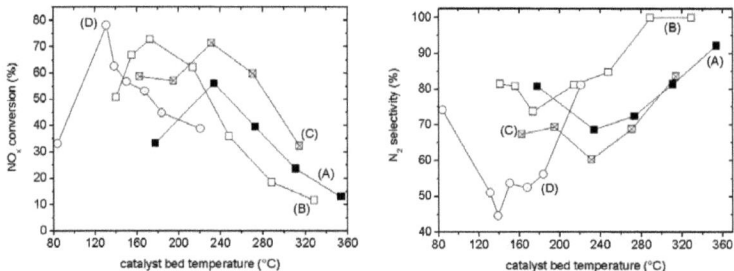

Figure 5.56 NO$_x$-reduction (left) and N$_2$-selectivity (right) during the NO$_x$-reduction under lean conditions with; LaCe$_{(0.05)}$Fe-Pd (A), La$_{0.6}$Ce$_{(0.4)}$Fe-Pd (B), BaTi-Pd (C), 150 mg of catalyst 1 wt. % Pt/SiO$_2$ (D). Feed composition: 720 ppm NO, 5 vol. % O$_2$, 1 vol. % H$_2$, 7.2 vol. % H$_2$O, 7.2 vol. % CO$_2$, He as balance. Reaction conditions: 300 mg of catalyst, at a W/F = 0.065 g$_{cat}$.s.ml^{-1}.

Increase in the Ce-content of the La-based catalyst resulted in a slight improvement of the NO$_x$-conversion. Fig 5.57 displays the NO$_x$-conversion observed in the presence of water vapour at 170°C with the catalyst LaCe$_{(0.4)}$Fe-Pd. A maximum NO$_x$-conversion of 73 % was found with this catalyst at lower temperatures compared to the other samples. The observed improved catalytic behaviour may be related to the presence of segregated CeO$_2$ in the catalyst as confirmed by XRD. Alternatively, the improved catalytic activity may be due to the presence of highly homogeneous mixture of oxides. The catalyst LaCe$_{(0.4)}$Fe-Pd showed the lowest N$_2$O rate formation relative to N$_2$ in these series. Moreover, the maximum N$_2$O selectivity with 26.2 % corresponds to the maximum NO$_x$-conversion (73 %).

For comparison purposes, a platinum supported catalyst (1.0 wt. % Pt/SiO$_2$) provided by CenTACat was also tested under the same reaction conditions. Fig. 5.56 displays the NO$_x$ performance (left graphic) and N$_2$-selectivity (right graphic) of the 1 wt. % Pt/SiO$_2$ catalyst. The maximum NO$_x$-conversion with this catalyst was 78 % at 130°C. The Pt-supported catalyst showed higher reactivity at much lower temperatures compared to the perovskite based catalysts. However, the observed NO$_x$-conversion was at a relatively narrow temperature window (< 150°C). These results are in good agreement with the results of similar catalytic tests previously reported by Burch et al [4]. The perovskite based catalysts presented here showed better N$_2$-selectivities compared with the platinum supported catalyst 1.0 wt. % Pt/SiO$_2$.

In order to understand the N$_2$O mechanistic formation on the perovskite catalysts and to find out if the palladium-doped catalysts could decompose N$_2$O under the given reaction conditions, NO was replaced in the feed by 150 ppm of N$_2$O (± 6ppm). The N$_2$O concentration was only slightly affected during the reaction between 100°C and 400°C. This experiment demonstrates that once the N$_2$O was formed during the H$_2$-SCR of NO$_x$ over the perovskite BaTi-Pd, it hardly participates in any other reactions. Fig. 5.57 demonstrates that the N$_2$O outlet concentration remain unaffected during the experiment. The N$_2$O concentration reported in this graphic was calculated after numerical integration of the IR intensity bands at 2202 and 2236 cm^{-1} at each run. IR-spectra were recorded every five minutes during the reaction at the given temperature program. The arrows in the Figure indicate the points for temperature increase during heating from one temperature to another. The table on the right side in Fig. 5.57 shows the catalyst bed temperature measured at each run. These results indicated that the catalyst BaTi-Pd-500°C hardly decomposes N_2O at temperatures between 100°C and 400°C under the given reaction conditions.

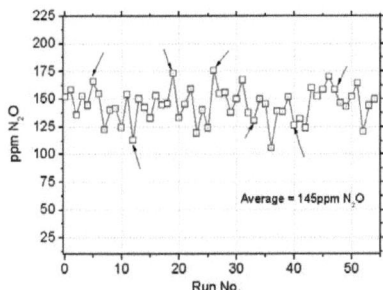

IR-spectra (Run no).	Temp. Cat. (°C)
0-5	112
7-12	136
14-19	155
21-26	171
28-33	209
35-40	255
42-47	299
49-54	344

Figure 5.57 N_2O outlet concentrations during the reaction over the catalysts BaTi-Pd calcined in air at 500°C/3h. Reaction conditions: 300 mg of catalyst, at a W/F = 0.065 $g_{cat}.s.ml^{-1}$. Feed composition: 150 ppm N_2O, 5 vol. % O_2, 1vol % H_2, 7.2 vol.% H_2O, 7.2 vol. % CO_2, He as balance.

In an attempt to test if the NO_2 molecule had been formed and consumed (i.e. a sort of intermediate) during the H_2-SCR of NO_x over the BaTi-Pd catalyst, NO was replaced by NO_2 as reactant keeping the concentration of the other gases constant. The presence of NO_2 would improve the NO conversion values similar to the NO_x conversion measured over Pd/TiO_2 when using NO_2 as a reactant. There was no improvement in the NO conversion at all, but NO_2 was chemisorbed on the catalyst surface, namely over the barium ions. The profile in Figure 5.58 shows that the NO_2 chemisorption takes place at the temperature range between 180°C and 260°C. Then the NO_2 outlet concentration starts to increase at 313°C and ends at 359°C indicating desorption of relatively high volumes of NO_2. For comparison, the NO outlet concentration is plotted against temperature and time on stream for the catalyst BaTi-Pd-500°C during the H_2-SCR of NO_x (Fig. 5.59).

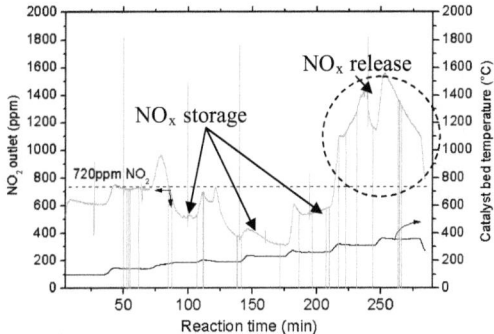

Figure 5.58 NO_2 as a reactant during the H_2-SCR of NO_x over BaTi-Pd calcined in air at 500°C/3h. Reaction conditions: 720 ppm NO_2, 5 % O_2, 1 % H_2, 7.2 % H_2O, 7.2 % CO_2 and He as balance, TFR = 276 ml.min^{-1} and 150 mg of catalyst.

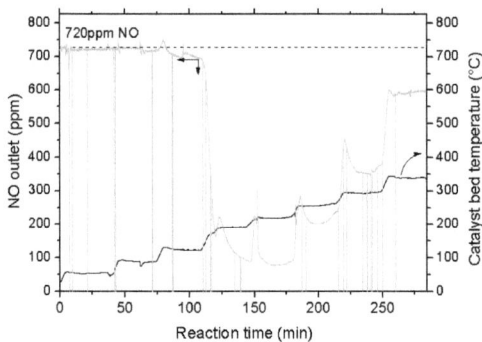

Figure 5.59 NO as a reactant during the H_2-SCR of NO_x over BaTi-Pd calcined in air at 500°C/3h. Reaction conditions: 720 ppm NO, 5 % O_2, 1 % H_2, 7.2 % H_2O, 7.2 % CO_2 and He as balance, TFR = 276 ml.min^{-1} and 150 mg of catalyst.

5.3.3 NO_x-reduction with H_2 under lean conditions in the presence of CO

Lanthanum based catalysts

The results of the present study show that the presence of CO has a negative impact, mainly on the reactivity of the tested catalysts. Fig. 5.60 displays the corresponding catalytic activity during the H_2-NO_x-SCR reaction under lean conditions with CO in the gas stream. The NO_x-conversion for all the cerium free catalysts was considerably reduced with CO in the reactant mixture. The decrease in Co-content in the catalyst LaFeCo$_{(0.475)}$-Pd to LaFeCo$_{(0.3)}$-Pd resulted in a decrease of the NO_x reduction ability, particularly between 210°C and 340°C. Whereas the substitution of the La-site with cerium resulted in a remarkable improvement of the catalytic activity between 180°C and 285°C even in the presence of CO (see Fig. 5.60). Further increase in the catalyst bed temperature resulted in a decrease in NO_x-conversion.

Fig. 5.60 displays the NO_x-reduction performance of the catalyst LaCe$_{(0.05)}$Fe-Pd under lean conditions with CO in the feed. A maximum of 62 % of NO_x-conversion was observed between 220°C and 250°C under the given reaction conditions. Similar NO_x-conversion profile was observed for the catalyst with cobalt, LaCe$_{(0.05)}$FeCo$_{(0.3)}$-Pd. These results indicate that the tetravalent cerium Ce^{4+} atoms positively affect the NO_x-reduction performance of the catalyst. The degree of substitution in the lanthanum-site of the perovskite structure with cerium plays an important role during the NO_x reduction as demonstrated here. The catalyst with small substitution of cerium at the A-site of the perovskite LaCe$_{(0.05)}$Fe-Pd display the highest NO_x-reduction performance between 220°C and 290°C even in the presence of CO (see curve C) for catalyst LaCe$_{(0.05)}$FeCo$_{(0.3)}$-Pd and curve (D) for the catalyst LaCe$_{(0.05)}$Fe-Pd in Fig. 5.60).

Fig. 5.61 shows the N_2-selectivity of the perovskite based catalysts tested during the NO_x-reduction including CO in the feed. In the series of the cobalt substituted catalysts, the perovskite LaFeCo$_{(0.475)}$-Pd (Fe/Co ratio = 1) shows the best catalytic performance with 54 % N_2–selectivity at 180°C and 46 % N_2-selectivity at 210°C and this remains almost constant up to about 290°C (left graphic in Fig. 5.61). The increase in the Fe/Co ratio from 1 to 2.17 in the perovskite LaFeCo$_{(0.3)}$-Pd leads to no improvement in the NO_x-reduction performance (left graphic in Fig. 5.60). However, lower N_2-formation rates were observed. A maximum N_2-selectivity of 34 % was measured at 237°C with this catalyst. At higher temperatures, the N_2-selectivity decreases almost by 10 % at 340°C due to the relative high N_2O production at this temperature. Co-free lanthanum-based catalysts display interesting NO_x-reduction performance and N_2-selectivity. The catalyst $LaCe_{(0.05)}Fe$-Pd shows good catalytic activity between 220° and 285°C with a N_2-selectivity of 52 % at 220°C and 56 % at 285°C. At temperatures above 280°C, the N_2-selectivity with this catalyst remains at around 56 % constant up to 320°C.

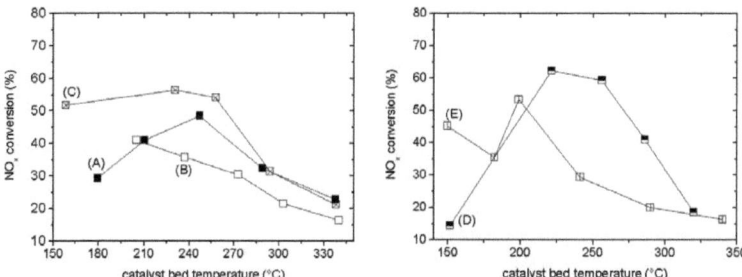

Figure 5.60 NO_x-conversion under lean conditions over $LaFeCo_{(0.475)}$-Pd (A), $LaFeCo_{(0.3)}$-Pd (B), $LaCe_{(0.05)}FeCo_{(0.3)}$-Pd (C), $LaCe_{(0.05)}Fe$-Pd (D), BaTi-Pd (E), all calcined at 900°C/3h. Reaction conditions: 300 mg of catalyst, at a W/F = 0.065 $g_{cat}.s.ml^{-1}$. Feed composition = 720 ppm NO, 5 vol. % O_2, 1vol. % H_2, 0.25 vol. % CO, He as balance.

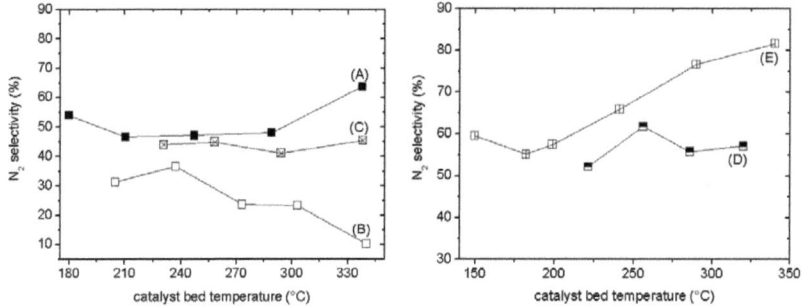

Figure 5.61 N_2-selectivity as a function of temperature with; $LaFeCo_{(0.475)}$-Pd (A), $LaFeCo_{(0.3)}$-Pd (B), $LaCe_{(0.05)}FeCo_{(0.3)}$-Pd (C), $LaCe_{(0.05)}Fe$-Pd (D), BaTi-Pd (E), all calcined in air at 900°C/3h. Reaction conditions: 300 mg of catalyst, at a W/F = 0.065 $g_{cat}.s.ml^{-1}$, feedstock: 720 ppm NO, 5 vol. % O_2, 1 vol. % H_2, 0.25 vol. % CO, He as balance.

Barium based catalyst

The NO_x conversion activity of the catalyst BaTi-Pd was noticeably affected by the presence of CO in the reaction gas mixture. The results in Fig. 5.60 display the decreased NO_x-reduction performance of the catalyst in the presence of CO. The NO_x-conversion resulting in N_2O formation follows the same trend as a function of the catalyst bed temperature. Two NO_x-conversion maxima are observed; one at 150°C and the other at 200°C. This behaviour clearly displays the impact of CO on the reactivity of the catalyst, implying that the CO directly oxidizes to form CO_2. The relative IR bands intensity of CO_2 to CO increases indicating the start of CO oxidation at 84°C (see Fig. 5.62). During the heating of the catalyst from 100°C to 150°C, the IR-bands for CO remain constant. As soon as the catalyst bed temperature reaches to about 150°C, CO oxidizes completely into CO_2. The simultaneous CO-oxidation and NO-reduction at the same catalytic Pd-sites may explain the decrease in NO_x-conversion at temperatures higher than 250°C. The catalyst BaTi-Pd shows similar N_2O production (approx. 70 ppm) in the presence of H_2O_{vapour} + CO_2 and with CO in the feed between 150°C and 200°C. At 240°C and above the N_2O formation decreases down to less than 30 ppm.

The catalyst BaTi-Pd shows a moderate NO_x-conversion with CO in the feed at temperatures below 200°C. As mentioned above, with this catalyst 60 % selectivity to N_2 was measured at 150°C. At 180°C the N_2-selectivity slightly decreases to 55 % and then increases again at 200°C and above (right graphic in Fig. 5.61).

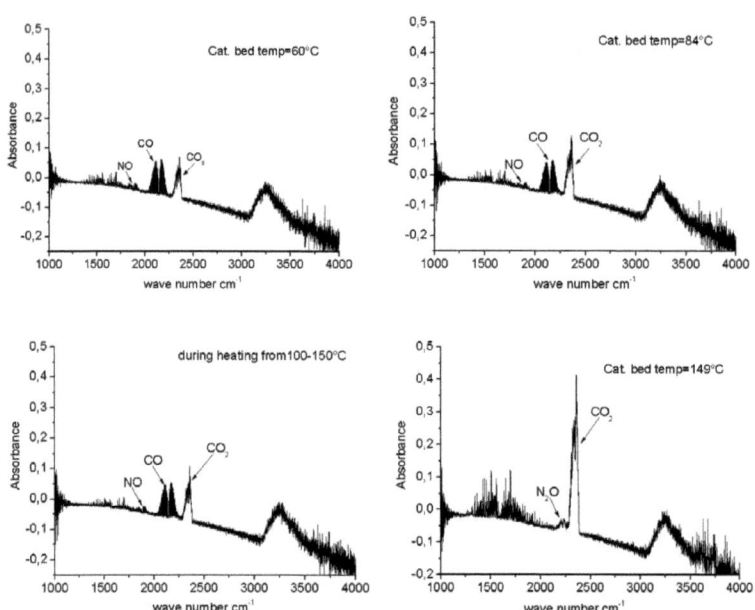

Figure 5.62 Gas phase IR-spectra during the NO_x-reduction with the catalyst BaTi-Pd calcined at 900°C/3h. Reaction conditions: 300 mg of catalyst, at a W/F = 0.065 $g_{cat}.s.ml^{-1}$, feedstock = 720 ppm NO, 5 vol. % O_2, 1 vol. % H_2, 0.25 vol. % CO, He as balance.

5.3.4 NO_x-reduction with propene under lean conditions (HC-SCR)

To determine the optimum conditions for the catalytic tests a first series of experiments were carried out by employing a catalyst BaTi-Pd-500°C with perovskite structure. Table 5.10 summarises the gas composition of the reaction mixture during these experiments. The gas mixture was prepared by mixing the corresponding volume of each gas in a small recipient (all test gases provided by Praxair). The oxygen concentration was intentionally kept at a lower level to avoid the oxidation of C_3H_6 which was expected to occur even at moderate temperatures. Similarly, the C_3H_6 concentration was varied from 250 ppm to 733 ppm to find out the optimum reducing agent concentration for these experiments. In a second series of experiments the oxygen concentration was varied by keeping the C_3H_6 concentration constant. In these catalytic tests the temperature was increased step by step (each step = 25°C) from 150 to 300°C. A dwell time of 30 min was enough at each step to reach the steady state in most cases. The NO-reduction began at 175°C and the maximum NO-conversion was reached between 200 and 225°C. At higher temperatures a decrease in NO-conversion is observed (see left graphic, Fig. 5.63). At 250°C and above, C_3H_6 oxidizes to form CO_2 and H_2O instead of reacting with the NO.

Table 5.10 First catalytic experiments during C_3H_6-SCR of NO_x carried out in a differential quartz tube micro-reactor.

Experiment	NO (ppm)	C_3H_6 (ppm)	O_2 (vol. %)	Balance	T (°C)
1	250	250-733	5	Argon	150-300
2	250	500	1-5	Argon	150-300

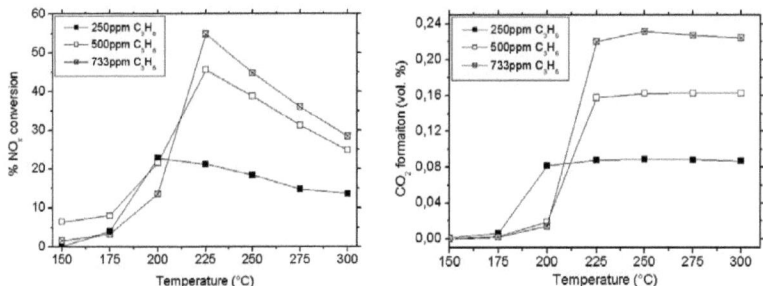

Figure 5.63 Conversion (left) and CO_2 formation (right) for the NO_x-reduction with 250 - 755 ppm of C_3H_6 as a reducing agent and 5 vol. % O_2 over a catalyst BaTi-Pd-500°C. Reaction conditions: 75 mg of catalyst pellets, at a W/F = 0.016 $g_{cat}.s.ml^{-1}$.

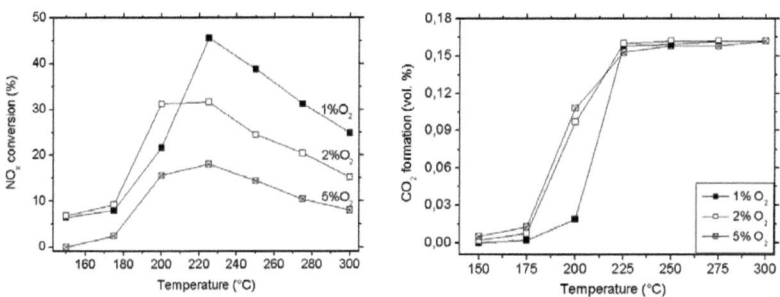

Figure 5.64 NO_x-conversion (left) and CO_2 formation (right) during the C_3H_6 –SCR of NO_x on 75mg of catalyst BaTi-Pd-500°C at a W/F = 0.016 $g_{cat}.s.ml^{-1}$. Feed composition = 250 ppm NO, 500 ppm C_3H_6, 1-5 vol. % O_2, Ar as balance.

Left graphic in Fig. 5.64 shows the effect of the increase in oxygen concentration on the NO_x-C_3H_6 reaction. A maximum of 45 % NO_x-conversion was reached at 250°C for a feed with 1 vol. % O_2, while an increase of the oxygen concentration to 5 % causes a decrease in the NO_x-conversion down to 18 % at the same temperature. A moderate NO_x-conversion was measured with 2 vol. % oxygen in the reaction gas mixture. Maximum NO_x-conversion was 31 % between 200° and 225°C under the given reaction conditions. Propene begins to oxidize at 200°C with 1 vol. % O_2 in the gas stream. Above 2 vol. % O_2 in the feed, the C_3H_6 begins to oxidize as early as 175°C. In all cases the C_3H_6 combustion was almost hundred percent completed at 225°C (see right graphic, Fig. 5.64).

NO_x-reduction with propene under lean conditions over perovskite catalysts

The reaction conditions for the catalytic tests were selected from the first experiments reported in this section. Table 5.11 summarises the catalysts tested for the NO_x-reduction with propene as a reducing agent under lean-burn conditions (see experimental section for detailed description of the test equipment). Two palladium substituted perovskites were studied and tested under the given reaction conditions. For comparison purposes palladium supported perovskites with the same palladium content were tested under the same reaction conditions. The perovskites without palladium were taken as a reference material.

Table 5.11 Perovskite based catalysts calcined at 700°C/3h tested for the NO_x-reduction with C_3H_6 under lean-burn conditions at SESAM (Institute of Materials Research of DLR).

Catalyst group A [a]	Catalyst group B [a]	Feed composition
LaFeCo$_{(0.3)}$-Pd	BaTi-Pd	250ppmNO,500ppmC_3H_6,1-5 vol. %O_2+Ar balance
Pd-LaFeCo$_{(0.35)}$	Pd-BaTi	
LaFeCo$_{(0.35)}$	BaTi	250ppmNO,500ppmC_3H_6,1-5 vol. %O_2+Ar balance
		250ppmNO,500ppmC_3H_6,1 vol. %O_2+Ar balance

[a] the formula implies the theoretical catalyst composition

Palladium supported catalyst Pd-LaFeCo$_{(0.35)}$

Fig. 5.65 shows the NO_x-conversion levels reached with the palladium supported catalyst Pd-LaFeCo$_{(0.35)}$ calcined in air at 700°C. A maximum NO_x-conversion of 35 % was reached at 250°C with 1 vol. % O_2 in the feed. The increase of the oxygen concentration to 5 vol. % in the reaction mixture shifted the maximum NO_x-conversion to lower temperatures. With this catalyst, a NO_x-conversion of 19 % is detected at 200°C when using 5 % O_2 in the feed. Similarly, the C_3H_6 light-off temperature is shifted to lower temperatures with increasing oxygen content in the reaction gas mixture as a result of the early C_3H_6 oxidation to CO_2 given in the right graphic of Fig. 5.65. T_{50} for C_3H_6 with 1 vol. % O_2 in the gas stream was 240°C, while T_{50} for C_3H_6 with 5 vol. % O_2 in the gas stream was 195°C. Propene (C_3H_6) was almost hundred percent converted to CO_2 at 200°C with 5 vol. % O_2 in the gas stream. Complete oxidation of C_3H_6 is reached at 250°C with 1 vol. % O_2 in the reaction gas mixture, while C_3H_6 was completely oxidized with 2 and 5 vol. % O_2 in the feed at 225°C.

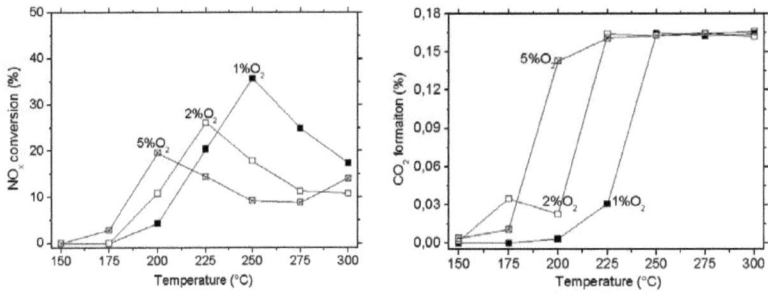

Figure 5.65 NO_x-conversion (left) and CO_2 formation (right) during the C_3H_6-SCR of NO_x with 1-5 vol. % O_2 on 2.5 wt. % Pd-LaFeCo$_{(0.35)}$-700°C catalyst. Reaction conditions: 75 mg of catalyst pellets, at a W/F = 0.016 $g_{cat}.s.ml^{-1}$.

Palladium substituted catalyst LaFeCo$_{(0.3)}$-Pd-700°C

The palladium substituted catalyst LaFeCo$_{(0.3)}$-Pd-700°C in turn shows interesting catalytic properties under the given reaction conditions. A maximum NO_x-conversion of 34 % was observed at 250°C with 1 vol. % O_2 in the gas stream (left graphic Fig. 5.66). The maximum NO_x-conversion was slightly shifted to lower temperature with 2 vol. % O_2 in the feed, a maximum NO_x-conversion of 28 % was obtained at 225°C. The increase of the oxygen content to 5 vol. % in the gas stream caused a shift of the maximum NO_x-conversion to higher temperatures than those observed with the palladium supported catalyst for the same reaction conditions. 27 % of NO_x-conversion was detected over the perovskite catalyst LaFeCo$_{(0.3)}$-Pd-700°C with 5 vol. % O_2 in the gas stream at 300°C, while the NO_x-conversion decreases at temperatures above 350°C (see Fig. 5.67). An experimental error can be ruled out since the catalytic test was carried out under the same reaction conditions three times, and very similar results were obtained in each case. Compared with the palladium supported catalyst, the C_3H_6 light-off temperature was shifted to higher temperatures; T_{50} = 210°C with 1 vol. % O_2, T_{50} = 218°C with 2 vol. % and T_{50} = 220°C with 5 vol. % O_2. Almost complete C_3H_6 oxidation was measured at approx. 225°C with 2 and 5 vol. % O_2 in the gas stream with the catalyst LaFeCo$_{(0.3)}$Pd-700°C, while C_3H_6 oxidation was hundred percent completed at 275°C for a reaction mixture having 1 vol. % O_2 in the feed.

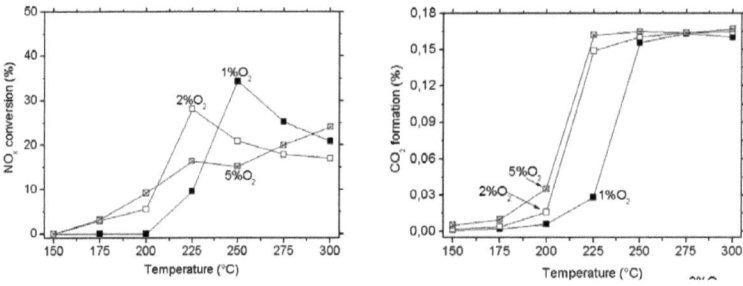

Figure 5.66 NO_x-conversion (left) and CO_2 formation (right) during the C_3H_6-SCR of NO_x with 1-5 vol. % O_2 on catalyst LaFeCo$_{(0.3)}$-Pd-700°C. Reaction conditions: 75 mg of catalyst pellets, at a W/F = 0.016 $g_{cat}.s.ml^{-1}$.

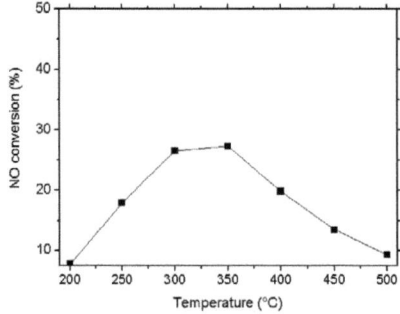

Figure 5.67 NO_x-conversion during the C_3H_6-SCR of NO_x with 5 vol. % O_2 on catalyst $LaFeCo_{(0.3)}$-Pd calcined at 700°C/3h. Reaction conditions: 75 mg of catalyst pellets, at a W/F = 0.016 $g_{cat}.s.ml^{-1}$.

Palladium supported catalyst Pd- BaTi-700°C

Palladium supported catalyst Pd-BaTi calcined in air at 700°C/3h (~ 2.5 Pd wt. %) showed in turn a maximum NO_x-conversion of 33 % at 225°C with 1 % O_2 in the feed. The maximum NO_x-conversion of 30 % was shifted down to 200°C over the catalyst Pd-BaTi-700°C with 2 vol. % O_2. Further increase in the oxygen concentration causes a decrease in the NO_x-conversion, yielding a maximum NO_x-conversion of 18 % at 200°C with 5 % O_2 in the feed (see left side in Fig. 5.68). In all cases the C_3H_6 conversion over the palladium supported catalysts Pd-BaTi-700°C is almost hundred percent completed at 225°C. The light-off temperature of propene is shifted to lower temperatures with the catalyst Pd-BaTi-700°C compared to the La-based catalysts. The T_{50} of propene for this catalyst was 212°C with 1 vol. % O_2 and ca. T_{50} = 190°C with 2 and 5 vol. % O_2 in the in the gas stream (right graphic in Fig. 5.68).

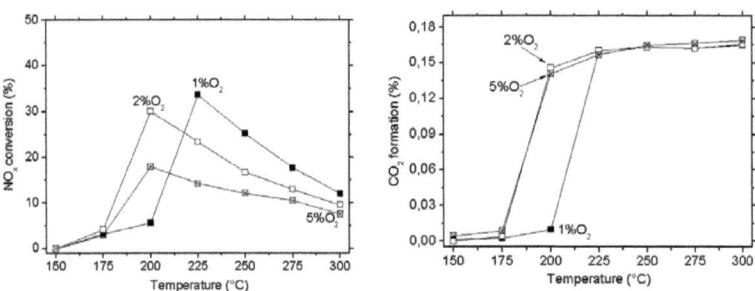

Figure 5.68 NO_x-conversion (left) and CO_2 formation (right) during the NO_x-C_3H_6 reaction with 1-5 vol. % O_2 on 2.5 wt. % catalyst Pd-BaTi-700°C. Reaction conditions: 75 mg of catalyst pellets, at a W/F = 0.016 $g_{cat}.s.ml^{-1}$.

Palladium substituted catalyst BaTi-Pd-700°C

The highest NO_x-conversion of 42 % was obtained at 225°C over the BaTi-Pd calcined in air at 700°C/3h catalyst with the addition of 1 vol. % O_2 in the feed. The maximum NO_x-conversion drastically decreases to 23 % at 225°C with 2 % O_2 in the gas stream. Similar NO_x-conversion values were obtained with 2 % and 5 % O_2 in the feed. Increase of oxygen concentration to 5 % results in solely to a shift in the maximum NO_x-conversion temperature to 200°C (left graphic, Fig. 5.69). The coincidence between the onset of the NO_x-conversion and the onset of the

CO_2 formation (or C_3H_6 oxidation) in each case is quit noticeable. The CO_2 formation profiles during the NO_x-reduction with increasing oxygen concentration in the feed are reported in the right graphic of Fig. 5.69.

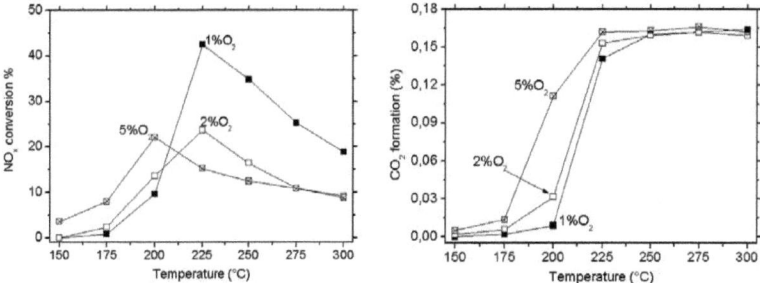

Figure 5.69 NO_x-conversion (left) and CO_2 formation (right) during the NO_x-C_3H_6 reaction with 1-5 vol. % O_2 on BaTi-Pd-700°C catalyst. Reaction conditions: 75 mg of catalyst pellets, at a W/F = 0.016 $g_{cat}.s.ml^{-1}$.

5.3.5 NO_x-reduction performance of the new catalytic coatings

As reported in section 5.2.2 the perovskite powder catalyst was coated on top of cordierite wash-coats provided by INTERKAT GmbH ®. The final average mass of perovskite on the coated wash-coats was approx. 3.72 g after calcination treatment. Considering ~ 2.5 wt. % of palladium in the perovskite catalyst, a Pd-content of 0.075 g can be estimated in the final wash-coat. This implies a palladium concentration of 58 g-Pd/ft^3 of catalyst with a wash-coat volume of 0.0366 l (or 0.00129252 ft^3). This volume includes the volume of the hollow channels of the wash-coat. In the literature, different palladium loadings were given depending on the catalyst application. 120g Pd/ft^3 was reported for close-couple converter catalysts, while 40 g-Pd/ft^3 of catalyst for under-floor converter (see p. 181 of [9]). In order to prove the catalytic properties of the catalytic converter coated with the perovskite LaFeCo$_{(0.3)}$-Pd developed in this study, C_3H_6-SCR of NO_x test under lean conditions is employed.

Table 5.12 summarises the number of reactions and feed compositions tested over the catalytic converter LaFeCo$_{(0.3)}$-Pd-cordierite. The light-off temperatures measured for the different reaction conditions are also reported in the Table 5.12. These catalytic tests were carried out at the facilities of the company INTERKAT GmbH ®. The NO-reduction started at 250°C and reached to a maximum NO-conversion of 20 % at 350°C by remaining constant up to 450°C (see Fig. 5.70). A slight decrease in the NO-conversion was detected with the increase of oxygen content in the feed from 1 vol. % to 5 vol. % O_2, while slightly lower light-off temperatures for C_3H_6 were recorded with this oxygen increment (307 -320°C). The presence of water vapour in the feed caused a positive effect on the NO-conversion over the catalytic converter LaFeCo$_{(0.3)}$-Pd-cordierite. The NO-conversion positively increased from 15 vol. % to 20 vol. % with 3.9 vol. % H_2O_{vapour} in the feed (see Appendix 2). On the other hand the light-off temperature of C_3H_6 was not substantially affected with the presence of water vapour as it can be seen from the results reported in Table 5.12. See appendix A2 for results of the tests of the catalytic converter (1, 3 and 4).

The catalytic coatings LaFeCo$_{(0.3)}$-Pd -EB-PVD reported in section 5.2.1 were tested at the facilities of the High Temperature and Functional Coatings of DLR's Institute of Materials

Research. The C_3H_6-SCR of NO over the catalytic coating LaFeCo$_{(0.3)}$-Pd -EB-PVD was carried out under the reaction conditions which were similar to the catalytic tests of the powder catalysts reported in the present work. No catalytic activity was observed during the C_3H_6-SCR of NO with 1 vol. % O_2 of the coating LaFeCo$_{(0.3)}$-Pd-EB-PVD. This result supports the importance of the geometric surface area of the catalyst for practical applications. As an example the actual geometric surface areas (GSA) of metallic substrates for automotive applications vary depending on the cell densities of the catalytic converter and this in turn affects its heat transfer properties. Automotive catalysts with high GSA's imply high cell densities (i.e. 6.8 m^2/l for a converter with 1600 cpsi) causing a negative effect in the heat capacity of the catalytic converter which causes longer cold-start periods. For comparison substrates with lower GSA's (i.e. 3.75 m^2/l for a converter with 400 cpsi) display better heat transfer capacities leading to shorter cold-start periods and lower NO emissions [9]. Higher geometric surface areas of the catalytic coatings LaFeCo$_{(0.3)}$-Pd-EB-PVD are needed in order to catalytically reduce NO with C_3H_6 under excess oxygen.

Table 5.12 Gas stream composition for catalytic test of the catalytic converter LaFeCo$_{(0.3)}$-Pd-cordierite at a SV = 60000 h^{-1} [b].

Reaction No	NO (ppm)	C_3H_6 (ppm)	O_2 (%) [c]	H_2O (%)	T_{50} (°C) [a]	NO_x-conv. (%)
1	510	515	1	0	320	15
2	510	520	1,1	3,9	318	14
3	510	520	5	0	307	11
4	460	495	4,7	3,5	312	---

[a] light-off temperature of C_3H_6
[b] see appendix 2 for results of reactions 1, 3 and 4
[c] nitrogen as balance gas

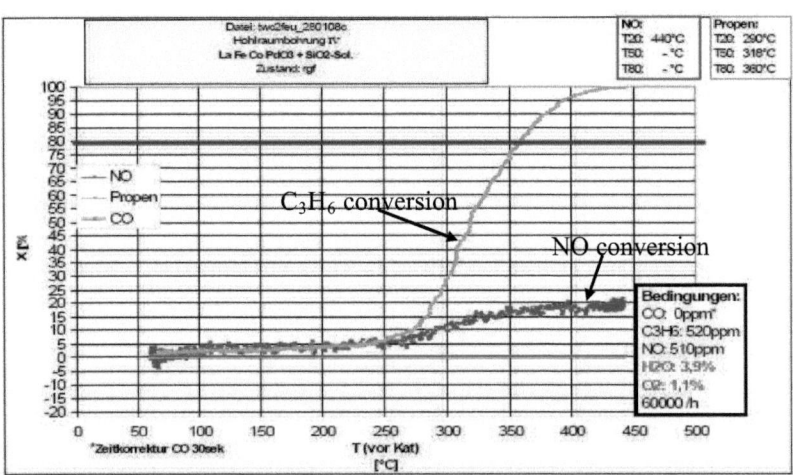

Figure 5.70 NO_x-conversion during the C_3H_6-SCR of NO_x over the catalytic converter LaFeCo$_{(0.3)}$-Pd-cordierite carried out at INTERKAT GmbH®. Feed composition: 510 ppm NO, 520 ppm C_3H_6, 1.1 vol. % O_2, 3.9 % H_2O_{vapour}, balance N_2, heating rate = 10 °K.min^{-1} at SV = 60 000 h^{-1}.

Chapter 6

6. Discussion

In this section the obtained results on crystal structure, phase(s) composition and properties of the synthesized perovskites are thoroughly discussed and compared with relevant studies. Emphasis is given to the knowledge and experience in the synthesis and characterization of the perovskites gained with this work at the research group of the high temperature and functional coatings division of DLR's Institute of Materials Research.

After the successful synthesis of the catalysts with perovskitic structure, first insights were gained with respect to the crystallization path of the lanthanum- and the barium-based perovskites up to 900°C by DSC analysis (in lanthanum-based perovskite), and the XRD-technique. The XRD diffractograms of the lanthanum- and the barium-based compounds and together with the structure refinement provided information about the perovskite purity. The oxidation state of palladium at the catalyst's surface, determined by means of XPS, gave new information about the surface composition of this kind of perovskitic structures. Particular attention was paid to the state of palladium in the synthesized perovskites analyzed by XRD, XPS and TEM after treatments at different temperatures under reducing and oxidizing atmospheres.

In this work, the lanthanum- and the barium-based perovskites were tested for the selective catalytic reduction of NO_x to simulate lean conditions under the presence of excess oxygen. A model gas mixture was selected using H_2 as a reducing agent (H_2-SCR NO_x) to prove the activity of the catalysts synthesized in the present work. The NO_x-conversion and N_2-selectivity results during the H_2-SCR NO_x reactions showed the potential of these perovskites to eliminate NO_x under lean conditions. The effect of the perovskite composition on the catalyst properties was studied with the purpose to establish composition-structure-properties relationships. The reaction conditions applied in this work, and the results of the catalytic activity tests during the H_2-SCR NO_x reaction, are compared with those of typical supported catalysts reported in the literature.

The selective NO_x-reduction capacity of the perovskite catalysts using C_3H_6 as a reductant was also tested. The NO_x-conversion values obtained here are compared with typical NO_x-conversions for different supported catalysts reported in the literature. Important to note is that no standard reaction conditions are found in the literature. The reaction conditions reported in the literature to test catalysts for C_3H_6–SCR NO_x and H_2-SCR NO_x reactions are varied and differ widely from each other. These differences complicate the objective discussion and comparison of the catalytic activity results. However, it is attempted to present an extense discussion.

First attempts to integrate the synthesized perovskites with the Electron Beam Physical Vapour Deposited (EB-PVD) PYSZ coatings were made in this study. The purpose of these experiments was to functionalize the EB-PVD PYSZ coatings with the catalytic active phase based on the synthesized lanthanum perovskite catalysts. Three variations of the direct impregnation coating method were carried out. The microstructure and phase composition of the EB-PVD PYSZ substrates coated with Pd-substituted lanthanum perovskite were analyzed. Similarly, the microstructural changes after a thermal ageing of the coating were investigated.

The lanthanum-based perovskite was successfully coated on cordierite substrates and first experiments were carried out to determine the C_3H_6-SCR NO_x capacity of this catalytic converter.

6.1 State of palladium in the La- and Ba- based perovskite catalysts

Two groups of palladium-substituted perovskites were synthesized: the La-based perovskites synthesized by a modified citrate route and Ba-based perovskites synthesized by employing the sol-gel method. Sections 4.1.1 & 4.1.2 yield a detailed description of the synthesis methods of these perovskites. These synthesis methods permit to obtain a single phase perovskite compounds. It is important to achieve single phase catalysts in order to enable sound structure-properties correlations of the perovskite catalysts. Palladium supported perovskites and Pd-free compounds were synthesized by applying the same synthesis method in each case. These Pd-free perovskites were used as a reference during the phase determination by means of XRD and XPS. The redox properties of the palladium-containing perovskites were studied with the XPS technique. A careful analysis of the palladium oxidation states in the different perovskites gives the information that explains the catalytic behaviour reported in the section 5.3.

In the present work one of the prime objectives was to demonstrate the incorporation of palladium ions into the frame lattice of the synthesized perovskite(s). Many reports in the literature indicate that palladium may be integrated into the structure of perovskite crystals. Tanaka et al [63] determined the state of palladium in a single phase lanthanum-based perovskite with the XPS technique. The binding energy of the palladium in the perovskites higher than that of bivalence palladium (Pd^{+2}) and lower than that of tetravalent palladium (Pd^{+4}) indicating the presence of palladium (i.e. Pd^{3+}) in the perovskite crystals as a solid solution [63]. On the other hand the perovskite with the composition $LaFe_{0.57}Co_{0.38}Pd_{0.05}O_3$ was apparently phase stable after calcination in air up to 900°C. The XRD results of the perovskite $LaFe_{0.57}Co_{0.38}Pd_{0.05}O_3$ reported later on, showed a single phase with perovskitic structure [73]. Particularly interesting is a small line close to the main reflection at about $2\theta = 34°$ which is associated with the perovskite structure. This reflection ($2\theta = 34°$) disappear upon reduction at 800°C in 10 vol. % H_2 in N_2, and then reappears after the re-oxidation at the same temperature [73]. The reflection at $2\theta = 34°$ may be due to the presence of PdO as reported by Zhou et al [97]. A perovskite with a similar composition to the last mentioned ($LaFe_{0.57}Co_{0.38}Pd_{0.05}O_3$) was synthesized by Zhou et al [97]. They synthesized a lanthanum-based perovskite ($LaFe_{0.77}Co_{0.17}Pd_{0.06}O_3$-900°C/5h) that show well defined superstructure reflections. According to the assumptions made by Zhou, in this perovskite the palladium ions were apparently stable after calcination in air up to 900°C for 5 h. However, controversial assumptions regarding the state of palladium are still found in the work of Zhou, and in general in the literature. So far there is no clear information about the thermodynamic stability of the perovskite crystal structure under oxidizing conditions at different temperatures. This is extremely important for practical applications since the catalyst(s) may be exposed to different temperatures ranging from room temperature up to 900°C or even higher depending on the catalyst application.

In the present work the citrate route was chosen for the synthesis of the La-based perovskite. A homogeneous mixture of the precursor elements is obtained with the citrate route (in other words a modified Pechini method). The precursor mass is amorphous on the synthesis and crystallizes upon calcination in air [98]. The phase(s) present in the calcined perovskite depend mainly on the synthesis method and on the Fe/Co ratio. The perovskite with the composition $LaFe_{0.8}Co_{0.2}O_3$ synthesized by applying the reactive grinding method fully crystallize as early as 600°C [99]. In contrast, the perovskites synthesized via the citrate route in this study crystallize at already 500°C on calcination in air. The perovskite $LaFeCo_{(0.3)}$-Pd with the orthorhombic structure began to crystallize at 600°C and was fully crystallized at around 660°C (see the DSC results in section 5.1.1). A single orthorhombic crystal phase was obtained after calcination at 700°C as confirmed by Rietveld refinement analysis. No other single or mixed oxides were detected by XRD. No reflections related to the presence of palladium oxide were found, suggesting to some extend the substitution of the B-site with palladium ions in the perovskite structure (ABO_3).

The Pd-substituted perovskite $LaFeCo_{(0.3)}$-Pd-700°C showed no other oxide species (i.e. PdO, Co_2O_3, or Fe_2O_3, etc). In contrast, the Pd-supported perovskite Pd-($LaFeCo_{(0.35)}$-900°C)-700°C displays the reflections corresponding to PdO (Fig. 6.1). After calcination at 750°C/3h no significant changes were observed in the main structure of the perovskite catalyst with composition $LaFeCo_{(0.3)}$-Pd. Regarding the presence of the reflexes $2\theta = 33.5°$ and 34° (JCPD 75-0584) palladium apparently segregates out of the crystal lattice as PdO. These results suggest a meta-stability of the perovskite structure having the composition $LaFeCo_{(0.3)}$-Pd at temperatures ranging from 500°C to 700°C. Under oxidation conditions at 750°C and higher temperatures, the palladium ions may migrate to form PdO as mentioned above. The lanthanum mixed oxides with palladium may decompose into the single oxide and metallic palladium at temperatures starting from 800°C depending on the La/Pd ratio [81]. The meta-stability of the system $LaFeCo_{(0.3)}$-Pd as determined may explain the phase(s) split and the PdO formation at temperatures higher than 700°C.

At 900°C the crystallinity of the sample increases as demonstrated clearly by the stronger reflections of the orthorhombic crystal ($LaFe_{1-x}Co_xO_3$) and tetragonal PdO (Fig. 5.2 in Chapter 5). The Pd-free perovskite $LaFeCo_{(0.35)}$ calcined at 900°C/3h presented no signals between $2\theta = 33.5°$ and 34.5°, thus confirming the phase purity of the perovskite synthesized with the modified citrate route. The reflections between $2\theta = 33.5°$ and 34.5° are related to the presence of other oxides such as $LaFeO_3$ and not to the presence of PdO in the cobalt free perovskites calcined at 800°C under oxidizing conditions [74], whereas the lines at $2\theta = 33.5°$ and 34° of the XRD patterns of the perovskite $LaFeCo_{(0.3)}$-Pd correspond to PdO tetragonal (see JCPD 75-0584). There is also a small reflection at $2\theta = 34.5°$ that corresponds to the orthorhombic perovskite $LaFeCo_{(0.35)}$. The XRD reflection at $2\theta = 34.5°$ was observed in the absence of palladium confirming this affirmation (see Fig. 5.3 in the Chapter 5).

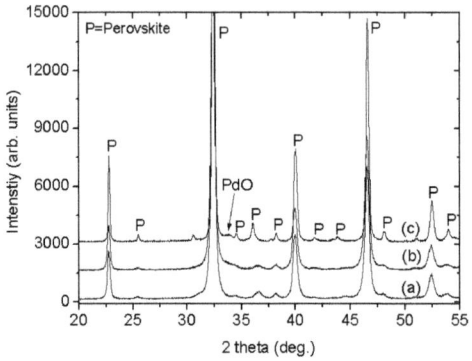

Figure 6.1 XRD-Patterns of (a) LaFeCo$_{(0.35)}$-700°C/3h, (b) LaFeCo$_{(0.3)}$-Pd-700°C/3h, (c) (Pd-LaFeCo$_{(0.35)}$-900°C)-700°C, data collection from 2θ = 20 to 98°, step size = 0.02°, counting time = 20 sec/step and 0.2 mm slit.

Small differences in the XRD patterns can be observed between the Pd-supported and the Pd-substituted perovskites after calcination at 900°C. The main XRD line of palladium as PdO in the perovskite LaFeCo$_{(0.3)}$-Pd-900°C is slightly shifted to higher 2θ angles when compared to the main XRD line of PdO from the perovskite Pd-LaFeCo$_{(0.35)}$-900°C (see Fig. 6.2). The substitution with palladium ions at the expense of cobalt produced almost no effect in the main signals of both La-based perovskites. Since the two samples were measured under the same experimental conditions, deviations in the XRD patterns caused by the instrument can be ruled out. In the Pd-supported perovskite Pd-LaFeCo$_{(0.35)}$-900°C the main peak of PdO was observed at 2θ = 33.8° which fits well to the tetragonal form of palladium oxide (JCPD 75-0584). This agrees well with the 2θ values which corresponds to the presence of PdO as reported in a supported perovskite LaFe$_{0.8}$Co$_{0.2}$O$_3$ [97]. Based on TEM images, the authors found PdO with particle sizes of 10 nm. The palladium agglomerates were classified into two categories; palladium particles fixed on well developed crystal planes of the crystalline LaFe$_{0.8}$Co$_{0.2}$O$_3$, and spherical particles weakly attached to the surface of the support. In the present study palladium agglomerates with sizes between 17 nm and 67 nm were observed for the Pd-supported perovskite Pd-LaFeCo$_{(0.35)}$-700°C.

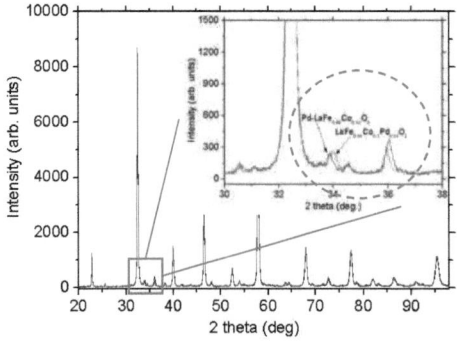

Figure 6.2 XRD-Patterns of the perovskites Pd-LaFeCo$_{(0.35)}$ and LaFeCo$_{(0.3)}$-Pd, both calcined in static air at 900°C/3h.

In contrast to the Pd-supported perovskite Pd-LaFeCo$_{(0.35)}$-900°C, the main XRD signal of palladium oxide in the Pd-substituted perovskite LaFeCo$_{(0.3)}$-Pd-900°C appeared slightly shifted to higher angles (2θ = 34°). The tetragonal PdO was slightly distorted in the Pd-substituted perovskite LaFeCo$_{(0.3)}$-Pd-900°C. These differences imply a strong interaction between palladium and the main perovskite in the catalyst LaFeCo$_{(0.3)}$-Pd. Such strong metal-support interactions have been solely assumed previously in the literature although no evidence was presented [97].

Analysis of the lanthanum-based perovskites by means of XPS provides information about the oxidation state of palladium in the catalyst's surface. The XPS binding energies (BE's) of palladium in different oxidation states reported in the literature are compared with the BE's of the perovskites analyzed here (see Table 6.1). The XPS line 337 eV of palladium in LaFe$_{0.95}$Pd$_{0.05}$O$_3$ lie between the corresponding XPS lines of Pd^{2+} and Pd^{4+} and indicate a single phase solid solution as affirmed by Uenishi et al. [74]. The BE at 337.7 eV of palladium in the perovskite BaCe$_x$Pd$_{(1-x)}$O$_{3-\delta}$ support this assumption [77]. However the BE's of palladium may be strongly affected by the presence of other host elements in the perovskite. Cobalt is an element that forms a solid solution with the lanthanum-based perovskite. The perovskite with the composition LaFe$_{0.57}$Co$_{0.38}$Pd$_{0.05}$O$_3$ was one of the first structures that present some kind of a crystal memory effect under redox environments. The XPS line at 337 eV for palladium, apparently intra-crystalline, is slightly shifted to lower values compared to the Co-free perovskite [63]. Assuming the complete incorporation of palladium in both perovskite crystal structures, the BE's shift indicates that the host elements may strongly affect the chemical environment of palladium. Previous studies observed no XPS signals from Pd 3d and implying highly dispersed palladium ions in the perovskite with the composition LaFe$_{0.77}$Co$_{0.17}$Pd$_{0.06}$O$_3$ [97].

In the present XPS study, BE of 336.5 eV for palladium in the perovskite LaFeCo$_{(0.3)}$-Pd indicate the presence of PdO (Pd^{2+}). PdO display the XPS line at 336.6 eV according to the literature [63, 74]. After calcination in air, PdO was found in the perovskites [LaFeCo$_{(0.3)}$-Pd and Pd-LaFeCo$_{(0.35)}$] synthesized in this study. Similarly, the palladium intra-crystalline was found at both samples. However, considerably less palladium was observed on the surface of LaFeCo$_{(0.3)}$-Pd than on Pd-LaFeCo$_{(0.35)}$ (see Table 5.5). Since the palladium content in both perovskites was calculated to be the same (approx. 0.05 mol %), a higher concentration of the intra-crystalline palladium is expected in the Pd-substituted perovskite LaFeCo$_{(0.3)}$-Pd than in the Pd-supported perovskite. Table 6.1 summarises the XPS Pd 3d BE's for the intra-crystalline palladium in perovskite lattices reported in the literature. Some controversies arise when looking carefully at the BE's of palladium in these compounds. The Pd-supported perovskite Pd-LaFe$_{0.8}$Co$_{0.2}$O$_3$ clearly displayed a Pd 3d BE of 336.7 eV which correspond to the palladium bivalent (Pd^{2+}) [97]. Other observations indicate that the intra-crystalline palladium display BE's of Pd 3d between 337.0 - 337.6 eV for the perovskites LaFe$_{0.57}$Co$_{0.38}$Pd$_{0.05}$O$_3$ [63] and LaFe$_{0.95}$Pd$_{0.05}$O$_3$ [74]. In this sense, the present study agrees well with the later reports. The Pd 3d BE at 336.4 and 336.5 eV measured at the perovskites LaFeCo$_{(0.3)}$-Pd and Pd-LaFeCo$_{(0.35)}$ correspond to the palladium bivalent, the BE at 337.7 and 337.5 eV are related with palladium in the lattice (i.e. The Pd-integrated state). It is quite obvious from the XPS spectra that there is a doublet with exceptionally high binding energy. It is not present in the initial impregnated perovskite Pd-LaFeCo$_{(0.35)}$ and survives (at least partially) the reduction treatment at 500°C in the Pd-substituted perovskite LaFeCo$_{(0.3)}$-Pd. This binding energy is higher than anything collected in the NIST database for Pd oxo-compounds. This might be a new indication of Pd in the crystal lattice where the BE may be influenced by the Madelung field of the trivalent La and Fe ions and in the Ba-based perovskite by tetravalent Ti ions.

Table 6.1 XPS lines of palladium in different perovskite based catalysts as-prepared (oxidized state).

Perovskite composition	Calcination temperature	Phase(s) [a]	Pd $3d_{5/2}$ [b]	Ref.
Pd-LaFe$_{0.8}$Co$_{0.2}$O$_3$	600°C/5h	Perovskite+PdO	336,7	[97]
LaFe$_{0.57}$Co$_{0.38}$Pd$_{0.05}$O$_3$	800°C/1h	Perovskite	337,0	[63]
LaFe$_{0.95}$Pd$_{0.05}$O$_3$	700°C/4h	LaFeO$_3$ + La$_2$O$_3$	337,6	[74]
PdO (Pd^{2+})	n. r.	PdO	336,6	[74]
0.1 wt. % Pd-LaCoO$_3$	600°C/5h	LaCoO$_3$, La$_2$O$_3$, Co$_3$O$_4$, Co	336,2	[100]
BaCe$_x$Pd$_{(1-x)}$O$_{3-\delta}$	1000°C/10h	Perovskite	337,7	[77]
LaFeCo$_{(0.3)}$-Pd	700°C/3h	Perovskite	337,3 & 336,4	This study
Pd-LaFeCo$_{(0.35)}$	700°C/3h	Perovskite	337,5 & 336,5	This study

[a] Determined by XRD
[b] BE in (eV) from the XPS measurements
n. r.: not reported
n. o.: not observed

The TEM observations on Pd-supported perovskite Pd-LaFeCo$_{(0.35)}$ indicate the presence of palladium particles between 15-20 nm. Palladium-rich particles/agglomerates of 5 to 10 nm and lower sizes were also found on the Pd-substituted perovskite LaFeCo$_{(0.3)}$-Pd. The EDX-analysis of the perovskite matrix indicates the presence of Pd in the perovskite matrix (i.e. Pd-ions in the perovskite lattice). The La-based perovskite LaFeCo$_{(0.3)}$-Pd reduced in hydrogen containing atmospheres yields 5 nm Pd-particles (Fig 5.14 in Chapter 5).

The redox behaviour of the Pd-substituted perovskite LaFeCo$_{(0.3)}$-Pd

Nishihata et al [12, 73], investigated the diffusion of Pd-ions and the alloy formation with cobalt at the perovskite LaFe$_{0.57}$Co$_{0.38}$Pd$_{0.05}$O$_3$ after the reduction in 10 vol. % H$_2$ + N$_2$ at 800°C/1h. About half of Co and around 10 % of the Fe ions were reported to be reduced to the metallic state, which possibly lead to the alloy formation between Co and or Fe with Pd. Similarly, the formation of an intermetallic alloy Pd$_3$La was proposed on Pd/Al$_2$O$_3$-La$_2$O$_3$ supported catalyst on the exposure to a NO + H$_2$ gas mixture at 700°C [101]. Besides the diffusion of Pd-ions in and out of the perovskite structure, the nano-scale alloy formation of Pd and Fe and/or Co, or even with La, are phenomena that cannot be completely ruled out.

No bulk palladium species were detectable in Pd supported on LaCoO$_3$ (Pd-LaCoO$_3$) after its calcination at 200°C suggesting a high degree of metal dispersion [100], but under model NO/H$_2$/O$_2$ gas mixtures palladium preserved its metallic character at the perovskite Pd-LaCoO$_3$ below 200°C (BE = 335.1 eV) [100]. The binding energy of the photopeak shifted toward higher energies (336.2 eV) after the air exposure of this catalyst at 300°C indicating re-oxidation of the metallic palladium. Moreover, another binding energy with an apparent maximum at 337.5 eV was detected indicating the presence of two different oxidised palladium states at this temperature (300°C) The binding energy at 337.5 eV corresponds to an intermediate value between those reported for Pd^{4+} in PdO$_2$ and Pd^{2+} in PdO and observed solely after exposing to reaction conditions for 2 h at 500°C. This BE can be attributed to the formation of the solid solution between palladium and the perovskite as reported by Uenishi [74].

According to the literature data, the oxidation state of the palladium ions dissolved in the perovskite lattice may be between the oxidation state of Pd^{2+} and Pd^{4+} and correspond to a XPS binding energy of Pd 3d equal to 337 eV [63]. Palladium is completely reduced to its metallic form at 800°C. After the re-oxidation in air for 1 h, it is reported that palladium shows the initial XPS binding energy of Pd 3d equal to 337 eV implying a fully reincorporation (solid solution formation) of palladium ions into the crystal lattice [74]. Singh et al [91] reported slightly different binding energies of Pd 3d for palladium in the solid solution with the perovskite $BaCe_{0.90}Pd_{0.10}O_{3-\delta}$. The interesting thing is that the binding energy of Pd 3d of 337.7 eV was still observed even after reduction at 1000°C/1h in 5 % H_2 containing atmospheres suggesting that some part of the palladium ions still remain inside the crystal lattice [77]. If palladium is supported on La-based perovskite of the similar composition as layer, the XPS binding energy of Pd 3d of 336.7 eV is measured which agrees well with the binding energy of Pd^{2+} (or PdO) [97]. The perovskites synthesized in this study display the Pd 3d binding energies 336.4 and 337.3 eV after reduction suggesting the presence of a metallic and an intermetallic palladium on the surface, respectively. The relative surface concentration of the constituting elements is higher in the Pd-supported perovskite Pd-$LaFeCo_{(0.35)}$ than that in the Pd-substituted perovskite $LaFeCo_{(0.3)}$-Pd, which suggests that a higher concentration of the intermetallic Pd is present in the Pd-substituted perovskite $LaFeCo_{(0.3)}$-Pd than the Pd-supported one.

In this study, two slight variations in the lattice parameter of the perovskite $LaFeCo_{(0.3)}$-Pd has been observed upon reduction under the hydrogen containing atmospheres (see Table 5.3). These variations may be due to the diffusion of Pd-ions out of the crystal structure. On the other hand, not much difference in the lattice parameters can be detected after the reduction and the re-oxidation implying that the lattice parameters are non-reversible modified on reduction. This may be due to the conditions applied for reduction process. The catalyst was reduced in a N_2/H_2 (20:1) gas mixture at 600°C, whereas the re-oxidation was carried out in static air atmosphere of a furnace. It is likely that during these extreme conditions oxygen vacancies and other structural defects are created in the perovskite structure which maintains the charge neutrality. On the other hand it must be kept in mind that the XRD may not be able to detect the small changes due to the palladium diffusion in an out of the perovskitic crystals due to the resolution limits of the instrument.

The XPS analysis of the perovskites synthesized in this study showed that the Pd-supported perovskite Pd-$LaFeCo_{(0.35)}$ contains Pd° on reduction (4.2 vol. % H_2 in Ar) at 200°C as confirmed with the Pd 3d binding energy of 334.8 eV. The literature data report similar binding energies for Pd° from pure PdO [63, 73, 74]. The palladium intra-crystalline was also reduced at 200°C in this sample. Palladium and the perovskite may react to form a solid solution at the most outer perovskite surface layer on calcination in air at 700°C. Fig. 6.3 displays a schematic presentation to demonstrate the state of palladium in the Pd-supported perovskite Pd-$LaFeCo_{(0.35)}$. In this Figure shows, a palladium oxide particle or agglomerate may diffuse into the perovskite surface lattice. The concentration of palladium in the surface lattice may depend on the calcination history of the catalyst and the reaction conditions. Intra-crystalline palladium (bulk palladium) is readily reducible at already 200°C under atmospheres containing 4.2 vol. % H_2 (Section 5.1.3).

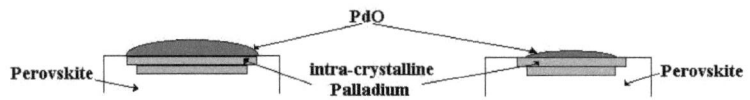

Figure 6.3 Schematic models of the Pd-supported perovskite Pd-LaFeCo$_{(0.35)}$ synthesized by the impregnation method.

The as-prepared perovskite LaFeCo$_{(0.3)}$-Pd showed the XPS BE line of Pd 3d at 336.4 eV that correspond to PdO (see Table 6.1). An additional BE at 337.3 eV observed in the XPS of this perovskite (in the as-prepared state) implies the presence of palladium in the crystal lattice [63, 74, 77]. Another characteristic is the Pd/La ratio of the perovskite LaFeCo$_{(0.3)}$-Pd is smaller compared to the Pd/La ratio of the Pd-substituted perovskite Pd-LaFeCo$_{(0.35)}$. Since the palladium content of both perovskites were adjusted to be the same on synthesis, the observed higher Pd-content in the Pd-substituted perovskite LaFeCo$_{(0.3)}$-Pd is an indication of the higher Pd-incorporation into the lattice or the perovskite matrix (see Table 5.5). Upon reduction, PdO and/or the surface lattice incorporated palladium species transform to metallic Pd.

On the contrary, report on the literature indicate that near-surface concentration of the perovskite BaCe$_x$Pd$_{1-x}$O$_{3-\delta}$ becomes poorer in palladium on reduction with 5 vol. % H$_2$ at 1000°C/1h [77]. In this case, it may be assumed that the Pd-concentration in the crystal structure of the perovskite will be higher implying a Pd-enrichment of the bulk as surface Pd-particles. The presence of a BE of Pd 3d at 337.7 eV indicates that intra-crystalline palladium is strongly bonded and remain in the crystal structure even though the catalysts BaCe$_x$Pd$_{1-x}$O$_{3-\delta}$ was treated in 5 vol. % H$_2$ at 1000°C/1h [77]. On re-oxidation for 1 h in O$_2$ at 1000°C of the perovskite BaCe$_x$Pd$_{1-x}$O$_{3-\delta}$, the palladium particles or agglomerates (~ 80 nm) were reabsorbed into the perovskitic crystal lattice.

The redox behaviour of the Pd-substituted perovskite BaTi-Pd

Both the Ba-based perovskites investigated in this study Pd-BaTi and BaTi-Pd showed binding energies of Pd 3d at 337.8 eV in the as-synthesized state. This BE may be due to the presence of the intra-crystalline palladium [63, 71, 74]. Moreover, the asymmetry of the Pd 3d peaks observed in Figs. 5.32 and 5.33 indicates the presence of an additional ionic palladium state, possibly the intra-crystalline palladium. With the Pd-supported perovskite (Pd-BaTi), a very low binding energy of Pd 3d is observed at 335.8 eV observed which can not correspond to PdO. In fact, one would expect the presence of an oxidized Pd-state (e.g. PdO) yielding higher binding energies (see Table 6.1). Also the applied synthesis route with calcination at 700°C/3h in static air implied that the formation of PdO type species is probably.

The as-synthesized Pd-substituted perovskite BaTi-Pd displayed Pd 3d binding energies at 336.0 eV. This is surprising because higher binding energies are associated with the intra-crystalline palladium (see Table 6.1). Even PdO show higher binding energies (Pd 3d 336.6 eV) than the Pd 3d BE's observed in the perovskite BaTi-Pd. Only PdO$_3$ (Pd 3d$_{5/2}$ 336.2 eV) displays such a binding energies (see NIST XPS Database) [102]. The XPS of palladium in the barium-based perovskite BaCe$_{0.95}$Pd$_{0.05}$O$_{3-\delta}$ presents a binding energy of Pd 3d at 337.7 eV [77] which agrees well with the XPS results obtained in this work with the Pd-substituted lanthanum-based perovskites (see table 6.1). In this study, different Pd 3d binding energies were detected in Pd-supported BaTi-perovskite and Pd-substituted BaTi-perovskites. (see Table 5.6). On reduction of the Pd-substituted BaTi-perovskite (BaTi-Pd) a Pd 3d binding energy at 334.7 eV was observed indicating a complete reduction of the surface lattice Pd into the metallic Pd state.

6.2 Catalytic activity of the Pd-substituted perovskites towards the NO_x-reduction with H_2 under lean conditions

This section gives the comparison of the NO_x-conversion values of the Pd-doped perovskites synthesized in this work with the NO_x-performance of typical supported catalysts reported in the literature. Under H_2-SCR conditions the effect of the perovskite composition (powder catalyst) to the NO_x-reduction and N_2-selectivity are analyzed. The effects of $H_2O + CO_2$ mixtures and CO presence to the NO_x-reduction are also studied under these conditions (H_2-SCR of NO_x). The use of propene (C_3H_6) as an alternative reductant in the selective catalytic reduction of NO_x over powder catalysts based on the perovskitic structure was tested (C_3H_6-SCR of NO_x). Similarly, the catalytic performance of the lanthanum-based perovskite coated on cordierite substrates was tested for dry- and wet- SCR of NO_x with C_3H_6 as reductant.

6.2.1 Effect of the cobalt-content in the perovskite $LaFeCo_{(x)}$-Pd on NO_x-reduction

The cobalt-content in the perovskites $LaFeCo_{(0.475)}$-Pd, $LaFeCo_{(0.4)}$-Pd, $LaFeCo_{(0.3)}$-Pd was intentionally varied at the expense of iron in order to reduce the cobalt oxide content segregated from the perovskite. The perovskite with 50 mol. % cobalt (x = 0.475) at the B-site contain clearly the cobalt oxide Co_3O_4 (in Section 5.1). The decrease in the cobalt content in the perovskite resulted in a decrease of the cobalt oxide detected before increasing the purity of the perovskite $LaFeCo_{(0.3)}$-Pd. Cobalt oxide was not detected with XRD in this perovskite composition. Independent of the cobalt content, all perovskites display certain amounts of the tetragonal PdO upon calcination treatment under oxidizing conditions at 900°C.

During the reaction (H_2-SCR) both perovskites $LaFeCo_{(0.475)}$-Pd and $LaFeCo_{(0.3)}$-Pd displayed a good NO_x-reduction activity (see Section 5.3). Maximum NO_x-conversion values of about 79 % were observed between 180°C and 230°C with the catalyst $LaFeCo_{(0.475)}$-Pd. These NO_x-conversion values are slightly higher than those measured within the same temperature window with the catalyst $LaFeCo_{(0.3)}$-Pd which yielded a maximum NO_x-conversion of 74 % at 200°C. The N_2-selectivity was shifted to higher temperatures with the Co-content decrease of the perovskite $LaFe_{(0.95-x)}Co_xPd_{0.05}O_3$. However the N_2O-formation was not shifted to higher temperatures (see right graphic in Fig. 6.4). The catalyst $LaFeCo_{(0.3)}$-Pd produced slightly less N_2O between 160°C and 240°C than the catalyst $LaFeCo_{(0.475)}$-Pd. The low surface area and phase composition of the perovskite $LaFeCo_{(0.3)}$-Pd may explain the lower NO_x-conversions compared to the perovskite $LaFeCo_{(0.475)}$-Pd (see Table 6.2).

In this study, as the reaction conditions were kept constant, the catalyst compositions were varied to establish their effect on catalytic activity. Table 6.3 compares the reaction conditions selected in this work for the H_2-SCR of NO_x reaction with the typical reaction conditions employed to test the platinum and palladium supported catalysts reported in the literature. The reaction conditions found in the published work differ from the reaction conditions used in this study. The NO concentration in the literature varies from 0.05 vol. % to a maximum of 0.25 vol. %. The oxygen concentration is maintained in a range between 5 and 6 vol. %. In some cases, low concentrations of carbon monoxide (less than 0.3 %) and water vapour (5 to 7 vol. %) are also included. It is not very objective to combine all these differences in the reactant mixture with the catalyst composition (i.e. metal loading) and the mass to flow ratio in order to define the effects on the NO_x-conversion and N_2-selectivity. Similar reaction conditions to those used by the group of Breen et al (CenTACat) are applied in this study [103]. One difference is that they used n-octane as a main reducing agent(s) with additions of 0.72 vol. % of hydrogen as promoter during the NO_x-reduction under lean conditions over Ag/Al_2O_3 catalysts [104]. Addition of even small volumes of hydrogen (450 ppm H_2) produced a positive effect on the NO_x-reduction performance of the silver supported catalysts [105]. In the presence of hydrogen alone, the silver supported catalysts were not able to reduce NO_x.

Table 6.2 Physical properties of the perovskite catalysts $LaFeCo_{(0.475)}$-Pd and $LaFeCo_{(0.3)}$-Pd

Perovskite[a]	Phase(s)[b]	BET SSA[c] m^2/g	% Pd[d]	Lattice constants Orthorhombic perovskite[e]	Lattice constants Tetragonal PdO[e]
$LaFeCo_{(0.475)}$-Pd	Perovskite, PdO, Co_3O_4	2.113	2.708	a1 = 5.4720 a2 = 5.5108 a3 = 7.7568	b1 = 3.0411 b2 = 5.3354
$LaFeCo_{(0.3)}$-Pd	Perovskite, PdO	1.657	2.06	a1 = 5.5031 a2 = 5.5258 a3 = 7.7970	b1 = 3.0284 b2 = 5.3475

[a] Catalysts calcined at 900°C/3h under oxidizing conditions
[b] Detected by XRD
[c] SSA: Specific Surface Area
[d] as PdO
[e] Calculated after Rietveld refinement

The platinum supported catalysts have been extensively reported as very active materials towards the H_2-SCR of NO_x under relatively high partial pressures of oxygen [106]. Unfortunately, the platinum based catalysts have shown extremely low formation rates of nitrogen. High volumes of nitrous oxides (N_2O) are also produced during the NO_x-reduction under lean conditions with hydrogen as a reductant [4]. Table 6.3 lists the literature data regarding reaction gas composition, the maximum NO_x-conversion and N_2-selectivity obtained with some platinum and palladium supported catalysts as well as the data obtained with the catalysts tested in this study. Platinum is very active towards the H_2-SCR of NO_x particularly at temperatures below 200°C [106]. Depending on the reaction conditions, the platinum based catalysts can be active towards the NO_x- reduction and selective to nitrogen as reported by Costa et al [83]. They reported for instance a maximum NO_x-conversion of 75 % and 80 % selectivity to nitrogen. Although different support composition and platinum loading quantities were used, as shown in Table 6.3 three Pt-based catalysts yield similarly high NO_x-reduction and N_2-selectivity. Despite, the very high catalytic activity obtained with the platinum supported catalysts, it is shown that they deactivate highly in the presence of small volumes of carbon monoxide [106]. Under real driving conditions small volumes of carbon monoxide are always present, particularly in diesel exhausts. Therefore, in order to define the properties of the catalysts synthesized in this work, in the following section, the effects of CO, H_2O and CO_2 are also tested and discussed.

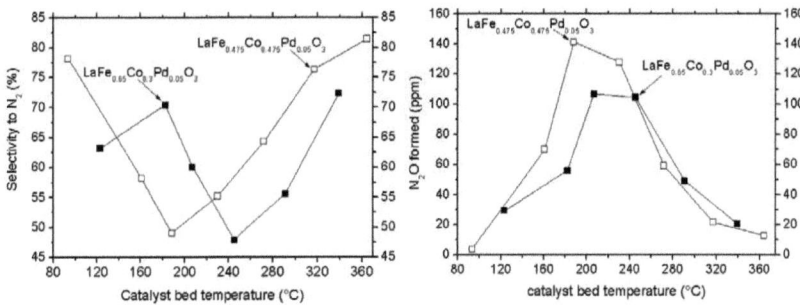

Figure 6.4 N_2-selectivity (left) and N_2O-formation in ppm (right) at the perovskite catalysts $LaFeCo_{(0.475)}$-Pd and $LaFeCo_{(0.3)}$-Pd. Reaction conditions: 720 ppm NO, 5 vol. % O_2, 1 vol. % H_2, He balance, TFR = 276 ml.min^{-1}, W/F = 0.065 $g_{cat}.s.ml^{-1}$.

In contrast to the platinum supported catalysts, the literature data shows that the palladium loaded catalysts are less active. Although, with the catalyst Pd-LaCoO$_3$ a maximum NO$_x$-conversion of 100 % was measured at 84°C the N$_2$-selectivity was only 64 % and the activity was limited to a narrow and low temperature window. It must be considered that this maximum was observed at a narrow temperature window (between 84° and 110°C). Moreover, this result was obtained at a extremely low space velocity, (e.g. 4000 h^{-1}). Below 80°C and above 110°C, the NO$_x$-conversion with this catalyst was extremely low and a major amount of N$_2$O was produced [75]. The properties of the catalyst Pd-LaCoO$_3$ were attributed to the possible modification of surface palladium species (i.e. the formation of oxidic noble metal species). There are some exceptional cases, where the Pd-supported catalysts show higher activity towards NO$_x$-reduction. Interestingly, it is reported that a palladium supported catalyst having V_2O_5 + TiO_2 mixed oxides as support can be very active towards the NO$_x$-reduction and be highly N$_2$-selective even at high space velocities [107]. The reason for these properties were explained to be due to the operation of a bi-functional reaction mechanism between the support and Pd. V_2O_5 which is a typical NH$_3$-SCR catalyst, may participate in the capture of reduced nitrogen species (e.g. NH$_3$ and NH$_4^+$) providing reaction sites for NO and O$_2$ to yield nitrogen. In analogous, in this study also the formation of NH$_3$ is observed during the catalytic tests of LaFeCo$_{(0.475)}$-Pd, however, this was only the case in the absence of hydrogen. The maximum NH$_3$-formation as detected with the IR band = 1121.8 cm^{-1} was at 240°C.

Table 6.3 Reaction conditions and NO_x-conversion during the H_2–SCR of NO_x.

Catalyst	NO (%)	O_2 (%)	H_2 (%)	H_2O_{vap} (%)	TFR (ml.min^{-1})[a]	W/F (g cat.s.ml^{-1})	Temp. of max. NO_x conv. (°C)	Max. NO_x conv. (%)	N_2-selectivity (%)	Ref.
0.1%Pt/La$_{0.5}$Ce$_{0.5}$MnO$_3$	0,25	5	1	---	100	0,09[c]	140	74	80	[5]
0.1%Pt/MgO-CeO$_2$	0,25	5	1	5	100	0,09[e]	150	95	85	[83]
0.3%Pt/La$_{0.7}$Sr$_{0.2}$Ce$_{0.1}$FeO$_3$	0,25	5	1	---	100	0,09[e]	150	83	93	[6]
1%Pt/Al$_2$O$_3$	0,05	6	0,2	5	200	0,03	95	75	Below 20	[4]
1%Pt/SiO$_2$	0,05	6	0,2	5	---	---	150	50	30	[4]
0.5%Pt/Al$_2$O$_3$	0,05	5	0,4	---	200	0,03	150	80	40	[106]
0.5%Pd/Al$_2$O$_3$	0,15	5	0,4	5	---	---[f]	275	9	6,5	[108]
1%Pd/LaCoO$_3$	0,5	3	0,5	---	---	---	84	100	63	[75]
1%Pd-5%V$_2$O$_5$/TiO$_2$	0,072	5	0,2[c]	---	200	0,03[g]	150	>95	95	[107]
LaFeCo$_{(0.3)}$-Pd	0,072	5	1[d]	7,2	276	0,065	222	58	57	This work
LaFeCo$_{(0.475)}$-Pd	0,072	5	1[d]	7,2	276	0,065	245	79	55	This work
LaCe$_{(0.05)}$FeCo$_{(0.3)}$-Pd	0,072	5	1[d]	7,2	276	0,065	241	62	60	This work
LaCe$_{(0.05)}$Fe-Pd	0,072	5	1[d]	7,2	276	0,065	234	56	68	This work
LaCe$_{(0.4)}$Fe-Pd	0,072	5	1[d]	7,2	276	0,065	172	73	74	This work
BaTi-Pd	0,072	5	1[d]	7,2	276	0,065	231	71	60	This work
Pt/SiO$_2$	0,072	5	1[d]	7,2	276	0,033	130	78	51	This work

[a] Balance He
[b] Depending on the catalyst density
[c] + 0,05 vol. % CO
[d] + 7,2 vol. % CO$_2$
[e] GHSV = 80000 h^{-1}
[f] GHSV = 4000 h^{-1}
[g] GHSV = 1,0*10^5 h^{-1}

If ammonia or similar species e.g. NH_4^+, NH_2 participate in the reaction mechanism during H_2-SCR of NO_x, this means that a reduced form of nitrogen is produced at the palladium surface and not at the support. The NO_x-reduction of the perovskite based catalysts are to be determined by the effect of the bulk structure of the perovskite into the electronic properties of palladium in the Pd-substituted perovskites $LaFeCo_{(x)}$-Pd. NO_x-conversion increases as the Fe/Co ratio in the Pd-substituted perovskite $LaFeCo_{(x)}$-Pd decreases at temperatures between 160° and 230°C. Although the NO_x-conversion increases as the Co-content increases in the $LaFeCo_{(0.475)}$-Pd catalyst as compared to the $LaFeCo_{(0.3)}$-Pd catalyst this is accompanied with a higher N_2O-production (see Fig. 6.4). This fact indicates that the decrease in the Co-content (or the increase of the Fe-content) in the La-based perovskite results in the formation of active sites that selectively reduce NO_x to nitrogen.

The effect of H_2O + CO_2 on the NO_x reduction capability of the catalysts $LaFeCo_{(0.475)}$-Pd and $LaFeCo_{(0.3)}$-Pd

The effect that H_2O_{vapour} has on the NO_x-conversion performance of platinum-supported catalysts is somewhat controversial. The presence of water vapour in the reaction gas mixture has a negative impact on the NO_x-conversion performance of platinum supported catalysts [4]. The presence of 5 vol. % H_2O_{vapour} caused a decrease of 10 % in the NO_x-conversion of 1 wt. % Pt/Al_2O_3 at temperatures above 130°C and an increment in the N_2O-production. A significant decrease in the N_2-production was measured over 1 wt. % Pt/SiO_2 at all temperatures. More recently Costa et al [83] reported the positive effect of 5 vol. % of H_2O_{vapour} on the NO_x-reduction and N_2-selectivity over 0.1 wt% $Pt/MgO-CeO_2$ catalyst. This catalyst did not deactivate at all under the reaction in presence of water vapour during 24 h at 200°C.

In this study the test carried out in the presence of water vapour and CO_2 over the Pd-substituted La-based perovskite catalyst ($LaFeCo_{(0.3)}$-Pd) yielded a temperature dependency of the reaction. Below 250°C the NO_x-reduction performance of the catalyst was decreased, whereas above 250°C, no change in the NO_x-conversion was observed (see the left side in Fig 6.5). At lower temperatures, it is likely that the water molecules compete with the NO-molecules in adsorption on the active sites, thus affecting the NO_x-performance and increasing the N_2O formation. As the temperature increases, less N_2O forms over the catalyst $LaFeCo_{(0.3)}$-Pd with water vapour and CO_2 in the feed (see right graph in Fig. 6.5). The water molecules dissociate on the catalyst surface and promote the formation of H_{ads} species increasing the N_{ads} concentration at the surface. This may occur according to the reaction NO_{ads} + H_{ads} → N_{ads} + OH_{ads} which was previously proposed by Dhainaut et al. [108]. Consequently the formation of molecular nitrogen increases as a result of the reaction between two adjacent chemisorbed nitrogen atoms as given by the reaction N_{ads} + N_{ads} → N_2 + 2* (* = adsorption site).

In contrast, the perovskite $LaFeCo_{(0.475)}$-Pd yields a clear improvement in the NO_x-conversion with water vapour in the feed (left side in Fig 6.6). The NO_x-conversion increases with the water vapour in the feed approx. 20 % more than the absolute NO_x-conversion under the dry conditions at 350°C. Also the formation of N_2O decreases with water vapour in the reaction mixture below 250°C remaining almost unchanged at higher temperatures. Water may dissociate on the catalyst surface facilitating the formation of molecular nitrogen and reducing the probability of the N_2O-formation. In this case higher NO_x-conversion is not coupled with higher formation rates of N_2O as in the 1 wt. % Pt/Al_2O_3 catalyst as reported by Burch et al [4]. They observed a higher rate of formation of N_2O with 5 vol. % of H_2O_{vapour} in the feed and gave no clear explanation for this (see Table 6.3).

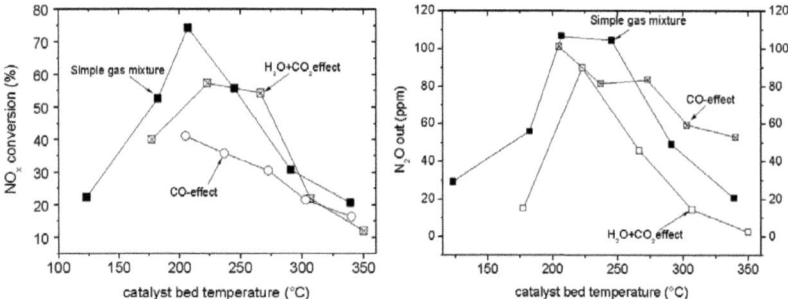

Figure 6.5 NO_x-conversion (left) and N_2O-formation in ppm (right) at the catalyst $LaFeCo_{(0.3)}$-Pd calcined in air at 900°C. Simple gas mixture: 750 ppm NO, 5 vol. % O_2, 1 vol. % H_2, He. CO effect: 750 ppm NO, 5 vol. % O_2, 1 vol. % H_2, 0.25% CO, He. H_2O + CO_2 effect: 750 ppm NO, 5 vol. % O_2, 1 vol. % H_2, 7.2 vol. % H_2O, 7.2 vol. % CO_2, He. (Data taken from figs 5.52, 5.55 & 5.60)

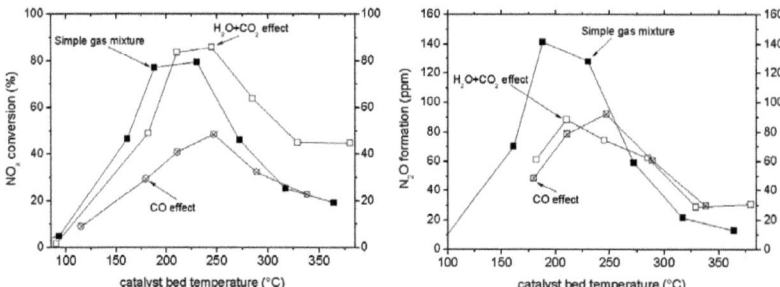

Figure 6.6 NO_x-conversion (left) and N_2O-formation in ppm (right) at the catalyst $LaFeCo_{(0.475)}$-Pd calcined in air at 900°C. Simple gas mixture: 750 ppm NO, 5 vol. % O_2, 1 vol. % H_2, He. CO effect: 750 ppm NO, 5 vol. % O_2, 1 vol. % H_2, 0.25% CO, He. H_2O + CO_2 effect: 750 ppm NO, 5 vol. % O_2, 1 vol. % H_2, 7.2 vol. % H_2O, 7.2 vol. % CO_2, He. (Data taken from figs 5.52, 5.55 & 5.60)

The effect of CO on the H_2-SCR reaction over the $LaFeCo_{(0.475)}$-Pd and $LaFeCo_{(0.3)}$-Pd catalysts

The effect of carbon monoxide on the activity and N_2-selectivity has been less studied [106]. Even small volumes of CO caused a strong decrease in the catalytic activity for H_2-SCR of NO_x over the platinum supported catalysts [106]. Carbon monoxide is considered as a poison for the perovskitic catalysts. The NO_x-reduction performance of both catalysts $LaFeCo_{(0.475)}$-Pd and $LaFeCo_{(0.3)}$-Pd decrease with CO in the feed. This observation agrees well with that reported in the literature [106, 107]. CO apparently reduces the rate of NO dissociation contributing to the loss of the NO_x-conversion and N_2-selectivity [106].

The NO_x-reduction and N_2-selectivity of the perovskite catalysts depend also on the Fe/Co ratio at the B-site of the perovskite $LaFeCo_{(x)}$-Pd. Table 6.4 summarises the volumes of N_2O formed during the H_2-SCR of NO_x in the presence of CO at the temperature range of the maximum

NO_x-conversion. The NO_x-conversion over the catalyst $LaFeCo_{(0.3)}$-Pd was positively affected in the presence of CO at temperatures below 250°C when compared to the NO_x-conversion of the catalyst $LaFeCo_{(0.475)}$-Pd. At 200°C less N_2O was formed over the catalyst $LaFeCo_{(0.475)}$-Pd than over the perovskite $LaFeCo_{(0.3)}$-Pd, whereas the catalyst $LaFeCo_{(0.475)}$-Pd produced slightly more N_2O than the catalyst $LaFeCo_{(0.3)}$-Pd at 250°C. At 300°C the catalyst $LaFeCo_{(0.475)}$-Pd produced less N_2O at higher NO_x-conversions than the perovskite $LaFeCo_{(0.3)}$-Pd. These tendencies indicate a complex N_2O-formation behaviour on the catalyst surface of both perovskites. The differences in the NO_x-conversion and the N_2O-formation rates can be attributed to the Co-content in these La-based perovskites ($LaFeCo_{(0.3)}$-Pd and $LaFeCo_{(0.475)}$-Pd). Less Co- or higher Fe-content in the perovskites may also modify the state of palladium. For instance the Pd-content in the perovskite $LaFeCo_{(0.475)}$-Pd was higher than the Pd-content in the perovskite $LaFeCo_{(0.3)}$-Pd as measured by fluorescence which indirectly indicates that less palladium segregates as oxide out of the crystal lattice of the perovskite $LaFeCo_{(0.3)}$-Pd (see Table 6.2).

Table 6.4 Catalytic properties of the catalysts $LaFeCo_{(0.3)}$-Pd and $LaFeCo_{(0.475)}$-Pd calcined at 900°C/3h in static air.

Catalyst	Catalyst bed temperature (°C)					
	200		250		300	
	NO_x conv. (%)	N_2O formed (ppm)	NO_x conv. (%)	N_2O formed (ppm)	NO_x conv. (%)	N_2O formed (ppm)
$LaFeCo_{(0.3)}$-Pd	41	101	55	81	21	59
$LaFeCo_{(0.475)}$-Pd	37	68	48	92	30	36

Reaction conditions: 300 mg catalyst, $W/F = 0.065\ g_{cat}.s.ml^{-1}$, 720 ppm NO, 5 % O_2, 1 % H_2, 0.25 % CO, He balance.

6.2.2 The role of cerium on the NO_x-conversion over the perovskite catalyst $LaCe_{(y)}$Fe-Pd

Cerium oxide has been reported to actively participate in the catalytic enhancement of the TWC's during the NO_x-reduction [20]. In the TWC, cerium oxide stores oxygen buffering the lean-rich swings in the exhaust gas composition during vehicle operation. This leads to a rich atmosphere and a better NO_x-reduction performance. The cerium oxygen storage capacity has been attributed to the combination of facile redox cycling between the trivalent and tetravalent oxidation states of the cerium ions [20]. Particularly, the perovskitic structure (ABO_3) incorporate cerium to some limit that lies below 0.05 mol % at the A-site improving the sintering resistance of the catalyst [63]. In addition, vacancies at the A-site arise due to the limited solid solution formation with cerium as tetravalent ions are produced at the B-site positively affecting the properties of the perovskite catalyst (i.e. the NO_x-conversion and N_2-electivity).

The XRD analysis of the perovskites synthesized in this work showed that the catalysts are composed of different crystal phases even with small cerium substitutions (see Table 5.5 in chapter 5). According to the literature the limits of mutual solubility of CeO_2 and the perovskite $LaCoO_3$ lies within x = 0.03 and 0.05 ($La_{1-x}Ce_xCoO_3$) [42]. Only small amounts of ceria were observed in the $La_{0.9}Ce_{0.1}CoO_3$ perovskite after calcination at 580°C. Higher calcination temperatures lead to a higher segregation of ceria as reported by Nitadori et al [109], indicating that CeO_2 segregation is temperature activated.

In this work, it was measured that the catalyst LaCe$_{(0.05)}$FeCo$_{(0.3)}$-Pd was less active than the perovskite LaCe$_{(0.05)}$Fe-Pd in terms of the NO$_x$-conversion. A better NO$_x$-reduction performance was observed with the catalyst LaCe$_{(0.05)}$Fe-Pd at temperatures between 260°C and 350°C compared with the NO$_x$-reduction performance of the catalyst LaCe$_{(0.05)}$FeCo$_{(0.3)}$-Pd as described in section 5.3.1 in chapter 5. The visible effect of the Co-addition was at the maximum NO$_x$-reduction temperature which is decreased from 227°C over the perovskite LaCe$_{(0.05)}$Fe-Pd to 182°C over the perovskite LaCe$_{(0.05)}$FeCo$_{(0.3)}$-Pd (see Table 6.3). Under a simple gas mixture containing NO, O$_2$, and H$_2$ diluted with He, the perovskite LaCe$_{(0.05)}$Fe-Pd displayed a slightly better selectivity to molecular nitrogen (see Fig. 6.7) than the later. However, the cobalt-free perovskite LaCe$_{(0.05)}$Fe-Pd produced 38.6 % more N$_2$O than the catalyst with cobalt LaCe$_{(0.05)}$FeCo$_{(0.3)}$-Pd at the temperature of the maximum NO$_x$-conversion (227°C). Considering that the formation of N$_2$ depends on the NO-dissociation according to the reaction N$_{ads}$ + N$_{ads}$ → N$_2$ as stated by Macleod et al [106], it might be feasible that the perovskite LaCe$_{(0.05)}$Fe-Pd easily dissociates NO particularly at temperatures between 250°C and 350°C. At the same time this catalyst facilitates the N$_2$O-formation supporting the reaction NO$_{ads}$ + N$_{ads}$ → N$_2$O + 2* [106] implying that both processes; the NO-dissociation and the NO-adsorption are improved. At the end, the overall reactions lead to a higher NO$_x$-reduction performance over the catalyst LaCe$_{(0.05)}$Fe-Pd.

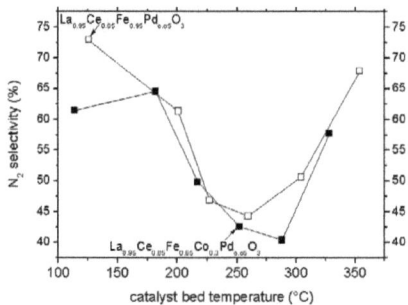

Figure 6.7 N$_2$-selectivity (left) and N$_2$O-formation in ppm (right) at the perovskite catalysts LaCe$_{(0.05)}$Fe-Pd and LaCe$_{(0.05)}$FeCo$_{(0.3)}$-Pd .Reaction conditions: 720 ppm NO, 5 vol. % O$_2$, 1 vol. % H$_2$, He balance, TFR = 276 ml.min^{-1}. (Data taken from fig 5.53)

The effect of H$_2$O$_{vapour}$ + CO$_2$ on the NO$_x$-reduction over the cerium based catalysts

The maximum NO$_x$-conversion over the catalyst LaCe$_{(0.05)}$FeCo$_{(0.3)}$-Pd was shifted to higher temperatures in the presence of H$_2$O$_{vapour}$ + CO$_2$ comparing to the reaction containing only NO, O$_2$ and H$_2$ (see left graph in Fig. 6.8). The maximum NO$_x$-conversions and the corresponding temperatures of the perovskite catalysts are summarized in Table 6.3. The effect that water vapour has on the NO-adsorption over some perovskites have been already documented [110]. The OH's formed on the support surface may react with the oxygen from the NO-molecules forming N$_{ads}$ + H$_2$O. N$_{ads}$ may react with NO forming the molecular nitrogen according to the reaction NO + N$_{ads}$ → N$_2$ + O$_{ads}$. Important to note is that the perovskite LaCe$_{(0.05)}$FeCo$_{(0.3)}$-Pd produced less N$_2$O in the presence of water vapour + CO$_2$ comparing to the reaction with only NO, O$_2$ and H$_2$ in the feed (see the right graph in Fig. 6.8).

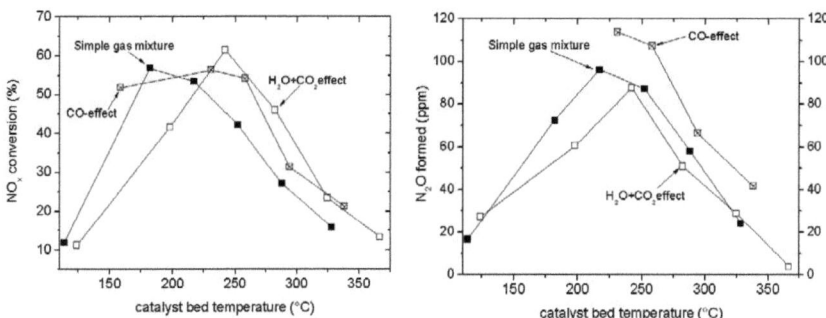

Figure 6.8 NO_x-conversion (left) and N_2O-formation in ppm (right) at the $LaCe_{(0.05)}FeCo_{(0.3)}$-Pd calcined in air at 900°C. Simple gas mixture: 720 ppm NO, 5 % O_2, 1 % H_2, and He balance. The $H_2O + CO_2$ effect: 720 ppm NO, 5 % O_2, 1 % H_2, 7.2 % H_2O_{vapour}, 7.2 % CO_2, He balance. The CO effect: 720 ppm NO, 5 % O_2, 1 % H_2, 0.25 % CO, He balance, 300 mg catalyst and W/F = 0.065 $g_{cat}.s.ml^{-1}$. (Data taken from figs 5.52, 5.55 & 5.60)

With the perovskite $LaCe_{(0.05)}Fe_{(0.95)}$-Pd a slightly negative effect on the NO_x-conversion was measured in the presence of water vapour + CO_2 in the feed (see the left graph in Fig. 6.9). But considerably less N_2O was produced over this catalyst in the presence of water vapour as compared to the catalytic test with the reaction mixture containing only NO, O_2, and H_2. An increase in the cerium concentration in the perovskite $LaCe_{(0.4)}$Fe-Pd shifted the maximum NO_x-conversion to approx. 73 % at lower temperatures (see Table 6.3). This catalyst showed high selectivity to nitrogen and a low formation of N_2O in the presence of water vapour (see Fig. 6.10). The NO_x-conversion and N_2-selectivity improvements are associated with the presence of high content of cerium oxide (see the XRD patterns in Fig. 5.12 reported in section 5). Cerium may affect synergistically the properties of palladium leading to an improved NO_x-reduction and N_2-selectivity. Previous work reported by Costa et al [6] showed that high NO_x-conversions to nitrogen can be obtained with the platinum supported catalyst 0.1 wt. % $Pt/La_{0.7}Sr_{0.2}Ce_{0.1}FeO_3$ (see Table 6.3). Based on the XRD analysis the authors found the $LaFeO_3$, $SrFeO_{3-x}$, Fe_2O_3 and CeO_2 oxide phases. Implying that single metal oxides may have an important effect on the NO_x-conversion and N_2-selectivity, the authors prepared the two platinum supported catalysts as reference materials, 0.1 wt. % Pt/Fe_2O_3 and 0.1 wt. % Pt/CeO_2. The Pt/CeO_2 system presented remarkably high values of the NO conversion in the 100° - 250°C range, similar to that obtained with the 0.1 wt. % $Pt/La_{0.7}Sr_{0.2}Ce_{0.1}FeO_3$ catalyst. However the Pt/CeO_2 catalyst showed comparably lower selectivities to nitrogen. In the case of the Pt/Fe_2O_3 catalyst, much lower NO-conversions and N_2-selectivities were obtained. All this suggest that the whole perovskite participate actively in the NO_x-reduction.

Cerium oxide participates in the reaction providing highly active sites for an oxygen chemisorption. In addition, cerium in solid solution with the perovskite may affect the electronic structure of palladium that might be introduced into the B-site of the perovskite. First, tetravalent cerium ions (Ce^{4+}) are partially dissolved in the Lanthanum site of the perovskite lattice whereas Lanthanum remains with its normal oxidation state La^{3+}. Considering the fully stabilization of Pd ions into the perovskite lattice of $LaCe_xFe$-Pd, this implies the co-existence of Pd^{4+} and Pd^{3+} ions in order to stabilize the valence of the whole perovskite. The presence of Pd^{4+} due to the substitution of the A-site with Ce^{4+}, and Pd^{3+} due to the presence of La^{3+}. Palladium species with different oxidation states in the perovskite may of course affect the overall NO_x-conversion performance of the catalyst. However, the presence of high Ce-content in the catalyst leads to a high content of Ce- and Fe- segregated oxides which also participate in the NO_x-reduction reaction in a positive or negative way as mentioned by Costa [6]. In the perovskite based catalyst single mixed oxides may provide the NO species (i.e. nitrates) that selectively reduce NO_x. It would be worth to demonstrate these hypotheses in further studies. In the present study the catalyst with the highest cerium content $LaCe_{(0.4)}$Fe-Pd displayed one of the best NO_x-reductions with the lowest N_2O-formation rates in the presence of water vapour and CO_2.

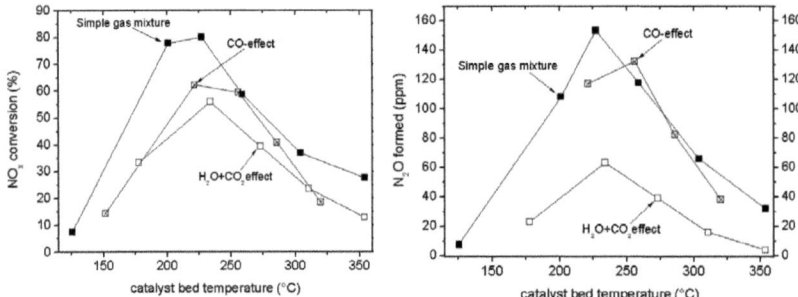

Figure 6.9 NO_x-conversion (left) and N_2O-formation in ppm (right) at the $LaCe_{(0.05)}$Fe-Pd calcined in air at 900°C. Simple gas mixture: 720 ppm NO, 5 % O_2, 1 % H_2, and He balance. The $H_2O + CO_2$ effect: 720 ppm NO, 5 % O_2, 1 % H_2, 7.2 % H_2O_{vapour}, 7.2 % CO_2, He balance. The CO effect: 720 ppm NO, 5 % O_2, 1 % H_2, 0.25 % CO, He balance. 300 mg catalyst, W/F = 0.065 $g_{cat}.s.ml^{-1}$. (Data taken from figs 5.52, 5.56 & 5.60)

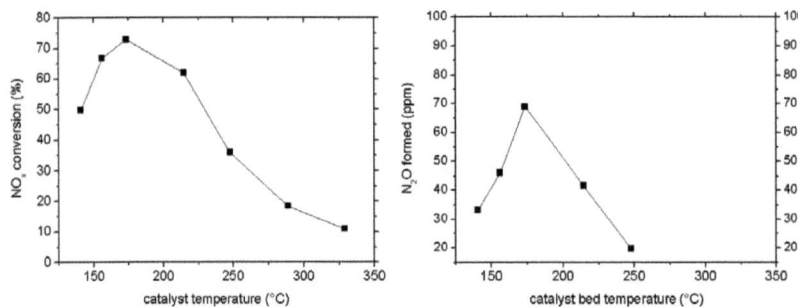

Figure 6.10 NO_x-conversion (left) and N_2O-formation in ppm (right) at the $LaCe_{(0.4)}$Fe-Pd calcined in air at 900°C. Reaction conditions: 720 ppm NO, 5 % O_2, 1 % H_2, 7.2 % H_2O_{vapour}, 7.2 % CO_2, He balance. 300 mg catalyst, W/F = 0.065 $g_{cat}.s.ml^{-1}$. (Data taken from fig 5.56)

The effect of CO on the H_2-SCR of NO_x over the $LaCe_{(0.05)}FeCo_{(x)}$-Pd catalyst(s)

In the presence of carbon monoxide, the NO_x-performance of the perovskite catalyst $LaCe_{(0.05)}FeCo_{(0.3)}$-Pd was slightly reduced between 150° and 250°C (see the left graph in Fig. 6.8). However the N_2O-formation was notably increased over this catalyst as seen from the right graph in Fig. 6.8. On the other side, the NO_x-performance of the catalyst $LaCe_{(0.05)}$Fe-Pd was negatively affected in the presence of CO at temperatures below 250°C (see the left graph in Fig. 6.9). The N_2O-production was not considerably affected over this catalyst by the presence of carbon monoxide (see the right graph in Fig.6.9). In both cases the maximum NO_x-conversion was observed between 200° and 250°C. Below 200°C only the La-based catalyst with cobalt in the crystal structure, $LaCe_{(0.05)}FeCo_{(0.3)}$-Pd, showed a good NO_x-performance with CO in the feed.

At 250°C the N_2O-formation is notably higher over the catalyst $LaCe_{(0.05)}$Fe-Pd than over the catalyst with cobalt in the perovskite lattice $LaCe_{0.05}FeCo_{(0.3)}$-Pd (see Table 6.5). Similar N_2O formation levels were measured over these two catalysts at 300°C, implying that cobalt had no effect on the N_2O-formation at least at this temperature with CO in the feed. During the catalytic test, the oxidation of CO may compete with the NO-dissociation at the same active sites causing formation of less N_{ads}–species thus showing a low NO_x-reduction and N_2-selectivities.

Table 6.5 NO_x-reduction and N_2O-formation of the catalyst(s) $LaCe_{(0.05)}FeCo_{(0.3)}$-Pd $LaCe_{(0.05)}$Fe-Pd calcined in static air at 900°C/3h.

	Catalyst bed temperature (°C)					
	200		250		300	
Catalyst	NO_x conv. (vol. %)	N_2O formed (ppm)	NO_x conv. (vol. %)	N_2O formed (ppm)	NO_x conv. (vol. %)	N_2O formed (ppm)
$LaCe_{(0.05)}FeCo_{(0.3)}$-Pd	54	----	55	110	30	62
$LaCe_{(0.05)}$Fe-Pd	47	----	60	130	32	65

Reaction conditions: 300 mg catalyst, TFR = 276 ml.min^{-1}, 720 ppm NO, 5 % O_2, 1 % H_2, 0.25 % CO, He balance.

6.2.3 The catalytic performance of the BaTi-Pd catalyst during H_2-SCR of NO_x

The barium based catalyst, designated by the formula BaTi-Pd, and showed a very good catalytic performance. This catalyst displayed the maximum NO_x-conversion of 73 % at about 170°C. At this temperature the catalyst BaTi-Pd was the most selective among the tested catalysts during the dry H_2-SCR of NO_x. Around 83 % of selectivity to nitrogen was registered at 170°C (see Table 6.3). Considering the same Pd-contents in La- and Ba-based catalysts, the effect of host perovskites becomes visible. Only the catalyst with the highest cerium loading $LaCe_{(0.4)}$Fe-Pd calcined in air at 900°C/3h showed a similar NO_x-performance and N_2-selectivities even in the presence of H_2O_{vapour} + CO_2 (see Fig. 5.56 in section 5.3). The presence of barium and titanium in the same structure play apparently a decisive role in the final catalytic properties of the Pd-substituted perovskite based catalysts.

The BaTi-Pd catalyst showed even better NO_x-conversions and N_2-selectivities than the platinum supported titanium oxide catalyst [24]. It is reported that the catalyst 1 wt. % Pt/TiO_2 displayed a maximum NO conversion of 50 % at 100°C with 40 % conversion to N_2O, the rest 10 % was converted to nitrogen, under gas mixtures containing 1000 ppm NO, 3000 ppm H_2, 5 vol. %

O_2 and 10 vol. % H_2O in He (W/F = 0.18 $g_{cat}.s.ml^{-1}$). Palladium supported on TiO_2 (1 wt. % Pt-TiO_2) reached 47 % of NO-conversion and 25.8 % conversion to N_2O under the same reaction conditions [24]. The NO-conversion was decreased down to almost zero at 200°C before reaching a second NO maximum conversion (44.9 %) and 17.5 % conversion to N_2O at 300°C [24].

In the work published by Ueda et al [24], the formation of NO_2 was proposed as indispensable for the selective reduction of NO_x on 1 wt. % Pd/TiO_2. The appearance of the two conversion maxima over this catalyst was proposed to be due to a change of the reaction paths between the direct reduction of NO by H_2 and the reduction of in-situ generated NO_2 by H_2. In fact, the reduction of NO_2 by H_2 took place faster than the NO + H_2 reaction suggesting the formation of NO_2 as an intermediate during the H_2-SCR of NO_x over the catalyst mentioned above [24]. The present work demonstrates contrary that NO_2 did not participate in the reaction mechanism as a main intermediate over the BaTi-Pd-500°C catalyst as propped by Ueda [24]. Instead NO_2 was selectively adsorbed on the perovskite surface under lean conditions, probably at the barium sites, and then desorbed without being reduced.

The effect of H_2O_{vapour} + CO_2 on NO_x-reduction over the BaTi-Pd based catalyst

The presence of H_2O_{vapour} + CO_2 caused a positive effect on the NO_x-reduction performance of the perovskite catalyst, BaTi-Pd, between 200° and 300°C. A maximum NO_x-conversion of 71.5 % was observed at 230°C even in the presence of water vapour (see the left graphic in Fig. 6.11). A slight decrease in the N_2O-production was measured in the presence of H_2O_{vapour} + CO_2 between 175° and 275°C over this catalyst (see the right graphic in Fig. 6.11). This perovskite, BaTi-Pd, was slightly worse in terms of N_2-selectivity than the catalyst $LaCe_{(0.4)}$Fe-Pd, which was the best catalyst tested in this study (see Table 6.3). Considering that the formation of N_2O requires the presence of NO_{ads} and N_{ads} over the catalyst surface according to a Langmuir-Hinshelwood mechanism as previously reported in the literature [108], it is plausible to indicate that water may compete with the NO-adsorption. This would also explain the NO_x conversion decrease in the presence of water vapour below 200°C. It is also possible that the water molecules may dissociate on the catalyst surface promoting the H_{ads} formation thus increasing the N_{ads}-concentration at the surface according to the reaction already mentioned before NO_{ads} + H_{ads} → N_{ads} + OH_{ads}. Consequently, the formation of molecular nitrogen would be increased. The shift in the temperature of the maximum NO_x-conversion from 170°C (without water vapour in the feed) to 230°C in the presence of water vapour may be an indication of the adsorption and dissociation of water molecules.

The NO_x-conversion values presented in this work are comparable with the NO_x-performance of the platinum supported catalysts reported by Costa et al [5]. The N_2-selectivity at the maximum NO_x-conversion under gas mixtures containing NO, O_2 and H_2 is similar to the selectivity obtained with the catalyst BaTi-Pd (see Table 6.3). The shift of the temperature of the maximum NO_x-conversion to higher values is an important difference between the catalytic performance of the palladium doped perovskite BaTi-Pd and the platinum supported catalyst reported by Costa [5]. In the present work a platinum supported catalyst (1 wt. % Pt/SiO_2) was tested under the same reaction conditions and used for comparison purposes. The catalyst 1 wt. % Pt/SiO_2 displayed a maximum 78 % of NO_x-conversion and 51 % of N_2-selectivity at 130°C. These results confirm the advantages of using Ba-based catalysts for H_2-SCR of NO_x reactions.

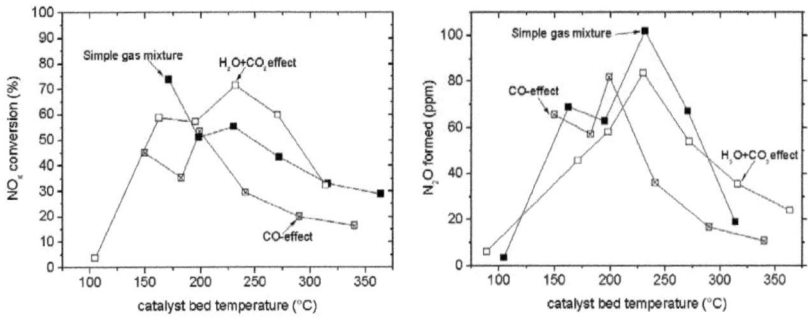

Figure 6.11 NO_x-conversion (left) and N_2O-formation in ppm (right) at the catalyst BaTi-Pd calcined in air at 900°C. Reaction conditions: 720 ppm NO, 5 % O_2, 1 % H_2, 7.2 % H_2O_{vapour}, 7.2 % CO_2, He balance, TFR = 276 $ml.min^{-1}$, W/F = 0.065 $g_{cat}.s.ml^{-1}$.

The effect of CO during the H_2-SCR of NO_x on the catalyst BaTi-Pd

The presence of CO in the feed caused a decrease in the NO_x-conversion of the catalyst BaTi-Pd. At 200°C and below, there is no clear tendency in the NO_x-conversion activity in the presence of CO. Above this temperature the NO_x-reduction behaviour decreases with CO in the feed (see the left graphic Fig. 6.11). In terms of the N_2O-production a complex behaviour was observed below 200°C over the perovskite BaTi-Pd. Above this temperature a relatively low concentration of N_2O was formed which indicates that CO may be oxidized at the same catalytic active sites. This would explain the decrease in the NO_x-conversion as well as the low N_2O-production. Under certain reaction conditions, i.e. high amount of catalyst and a relatively low space velocity, it has been proved that the perovskite support itself can be active towards the CO oxidation due to the presence of structural defects and anion vacancies [77]. The light-off temperature during the CO oxidation of 100 mg of the Ba-based perovskite ($BaCeO_3$) was T_{50} = 225°C under 1000 ppm CO, 10 vol. % O_2 in N_2 at a total flow rate of 50 $ml.min^{-1}$.

6.2.4 Catalytic activity of the catalysts BaTi-Pd and $LaFeCo_{(0.3)}$-Pd for C_3H_6-SCR of NO_x.

In this work propene was employed as a model reducing agent to eliminate NO_x under lean conditions over the perovskite-based catalysts containing palladium. The feed composition(s) are reported in Section 5.3.4. The propene concentration and the oxygen partial pressure were intentionally and systematically changed to study the catalytic performance of the synthesized perovskite-based catalysts. In this section the NO_x-reduction properties of the perovskite based catalysts doped with palladium $LaFeCo_{(0.3)}$-Pd and BaTi-Pd and calcined in static air at 700°C/3h are discussed and compared with data from the literature. The catalytic properties of the Pd-supported catalysts tested under the same reaction conditions are used as reference materials, (e.g. Pd-$LaFeCo_{(0.35)}$ and Pd-BaTi). The corresponding palladium free perovskites were also tested for the C_3H_6-SCR of NO_x reaction. The precious metal-free BaTi-perovskite ($BaTiO_3$) displayed no activity for this reaction and therefore it is not included in the discussion.

6.2.4.1. The C_3H_6-SCR of NO_x activity of the catalysts Pd-LaFeCo$_{(0.35)}$ and LaFeCo$_{(0.3)}$-Pd

The effect of the oxygen concentration during the C_3H_6-SCR of NO_x reaction

One of the main issues during the NO_x-reduction using hydrocarbons is the oxidation of the reducing agent. Increasing of the oxygen partial pressure during the reaction causes strong oxidation of the reducing agent. It is therefore a big challenge to selectively reduce NO_x to nitrogen with propene in the presence of excess oxygen. The perovskite based catalyst Pd-LaFeCo$_{(0.35)}$ displayed almost the same NO_x-conversion as the catalyst LaFeCo$_{(0.3)}$-Pd with 1 vol. % O_2 in the feed (see Table 6.6). The catalyst Pd-LaFeCo$_{(0.35)}$ displayed a decrease in the NO_x-conversion with increasing oxygen concentrations between 2 and 5 vol. % O_2. The higher the oxygen concentration in the feed, the more propene oxidizes instead of reacting with NO_x over this catalyst. Slightly a better NO_x-reduction performance was observed with the perovskites synthesized here compared to a typical supported catalyst such as Pd/Al$_2$O$_3$. This catalyst displayed one maximum of 25 % NO_x-conversion at 250°C with 5 vol. % O_2 in the feed (see table 6.6) [1]. The Pd-supported perovskite catalyst Pd-LaFeCo$_{(0.35)}$ (2 wt. % Pd) showed the maximum NO_x-conversion at 200°C. Generally, this difference in the NO_x-conversion can be associated with the metal loading. Increasing metal loadings per gram of catalyst cause an increase in the activity and so that the oxidation of propene begins at progressively low temperatures, at least over the platinum supported catalysts. There is a parallel increase in the NO-reduction which means that this reaction also begins at lower temperatures as the platinum loading increases [1].

The maximum NO_x-reduction over the Pd-substituted catalyst LaFeCo$_{(0.3)}$-Pd synthesized in the present work was slightly affected by the increasing concentration of oxygen in the feed. A maximum 34.3 % of NO_x-reduction was recorded with 1 vol. % O_2 in the feed at 250°C. The maximum NO_x-conversion was slightly reduced to 28 % and shifted to lower temperatures (225°C) with 2 vol. % O_2 in the feed. At higher oxygen concentration (5 vol. % O_2), the maximum NO_x-conversion of 30 %was shifted to even higher temperatures (300°C) over the catalyst LaFeCo$_{(0.3)}$-Pd. Experimental error can be ruled out since the catalytic test with 5 vol. % O_2 was repeated three times and pretty similar NO_x-conversion results were obtained. One would expect an increased depletion of propene at higher oxygen partial pressures in the feed at lower temperatures as in the supported catalyst Pd-LaFeCo$_{(0.35)}$. The results observed may be due to the partial incorporation of palladium ions in the perovskitic crystal lattice. A kind of small size effect of the palladium particles may cause the change in the NO_x-reduction trend with the increasing oxygen partial pressure in the reaction mixture. Amorphous palladium strongly bond to the perovskite surface may play also an important role during the C_3H_6-SCR of NO_x reaction. The presence of Pd^{2+} could be proven on the catalyst LaFeCo$_{(0.3)}$-Pd calcined in air at 700°C with the XPS analysis which delivers the information from the particles surface. Palladium oxide was detected after calcination at temperatures between 700° and 900°C. The XRD of the catalyst after being calcined at 900°C proved that PdO has a stronger interaction with the Pd-substituted perovskite LaFeCo$_{0.3}$-Pd compared to the PdO on the Pd-supported perovskite Pd-LaFeCo$_{(0.35)}$ (see Fig. 6.2 in section 6). A slight shift to lower 2θ values of PdO in the Pd-substituted perovskite LaFeCo$_{(0.3)}$-Pd was observed. These differences may be related to the synthesis method(s) employed to prepare both catalysts (see the experimental section). The modifications of palladium caused the interaction with the perovskite may be responsible for the observed catalytic behaviour.

Table 6.6 Catalytic activity of perovskites calcined in static air at 700°C/3h for the C_3H_6-SCR of NO_x reaction

Catalyst	Oxygen concentration in the feed					
	1%O_2		2%O_2		5%O_2	
	Max NO_x conv (%)	Temp. of Max NO_x conv (°C)	Max NO_x conv (%)	Temp. of Max NO_x conv (°C)	Max NO_x conv (%)	Temp. of Max NO_x conv (°C)
Pd-LaFeCo$_{(0.35)}$ [a]	35.6	250	26.0	225	19.6	200
LaFeCo$_{(0.3)}$-Pd [a]	34.3	250	28.0	225	30.0	300
Pd-BaTi [a]	33.6	225	30.0	200	17.9	200
BaTi-Pd [a]	42.4	225	23.6	225	22.0	200
LaFeCo$_{(0.35)}$ [a]	6.4	300	----	----	----	----
1%Pd/Al$_2$O$_3$ [b] [1]	----	----	----	----	25.0	250
LaMnO$_3$ [c] [111]	----	----	----	----	19	400

[a] Reaction conditions: 250 ppm NO, 250 ppm C_3H_6, 1-5 vol. % O_2, Ar balance. TFR = 300 ml.min^{-1} and 75 mg of cat (4000 cm^3/min.g$_{cat}$), this work
[b] Reaction conditions: 500 ppm NO, 1000 ppm C_3H_6, 5 vol. % O_2, (2000 cm^3/min.g$_{cat}$)
[c] Reaction conditions: 1000 ppm NO, 500 ppm C_3H_6, 5 vol. %O_2 and 5 vol. % H_2O (833.3 cm^3/min.g$_{cat}$)

6.2.4.2. The C_3H_6-SCR of NO_x activity of the catalysts Pd-BaTi and BaTi-Pd

The effect of the oxygen concentration on the C_3H_6-SCR of NO_x reaction

Increase of the oxygen partial pressure caused a decrease in the NO_x-reduction of the palladium supported Ba-based perovskite catalyst Pd-BaTi. Similar to the properties observed for the lanthanum-based catalyst mentioned above, the increase in the oxygen concentration causes high oxidation rates over this catalyst Pd-BaTi. Propene strongly oxidizes instead of reacting with NO. The early formation of CO_2 with the increasing oxygen concentrations suggests a change in the oxidation rate of propene (see Fig. 5.60 in section 5.3.4). According to the mechanism proposed by Huuhtanen et al [112], the reduction of NO is initiated by the adsorption of NO and propene, and then adsorbed propene reacts with the surface oxygen to partially oxidized hydrocarbons ($C_xH_yO_z$). These compounds may react with NO or oxygen to organic nitrogen compounds, acetates, formates and isocyanates which were identified as possible intermediates for the NO_x-reduction with hydrocarbons [112]. The reaction between nitrates and isocyanates or cyanides produces then nitrous oxide, nitrogen and oxygen. It is probably that the increasing oxygen content in the feed leads to the higher oxidation rates of propene and the reaction intermediates occur over the Pd-supported catalyst Pd-BaTi.

High oxygen concentrations in the feed caused also a decrease in the NO_x-reduction when using C_3H_6 as a reductant over the catalyst BaTi-Pd calcined in air at 700°C. The NO_x-reduction is decreased to absolute 20 % with the increase of the oxygen concentration from 1 to 2 vol. %. Propene is easily oxidized in the presence of high concentrations of oxygen. The onset of CO_2 formation observed at lower temperatures with 5 vol. % O_2 in the feed indicates the early oxidation of propene with the increasing concentration of oxygen in the reaction gas mixture (see Fig. 5.61 in section 5.3.4). The palladium free perovskite BaTi calcined in air at 700°C showed no activity for NO_x-reduction with propene as a reductant with 1 vol. % O_2 in the feed indicating that the host perovskite does not directly participate in the reaction under the mentioned reaction conditions. One example was published by Buciuman et al [111]; they reported a maximum 19 % of NO_x-conversion to nitrogen at 400°C over a perovskite having $LaMnO_3$ (see Table 6.6 for reaction conditions). The ratio between volume and the catalyst mass was approx. 5 times smaller over the perovskites synthesized by Buciuman et al [111] than the ratio(s) employed here in our experiments. Thus other than the possibly composition effect (i.e. Mn^{2+}), it can be implied that at low space velocity the Pd-free perovskites (the barium based perovskites) may be able to reduce NO_x.

It can be attributed that the NO_x-reduction ability of the barium based catalysts strongly depends on the palladium state. These two perovskites, the Pd-substituted and the Pd-supported catalysts, BaTi-Pd and Pd-BaTi respectively, both calcined in air clearly show different catalytic performances during the C_3H_6-SCR of NO_x reaction. Since the host Pd-free perovskite $BaTiO_3$ did not show any catalytic activity for C_3H_6-SCR of NO_x it is plausible to argue that changes in the performance of the catalyst are associated with the state of palladium, i.e. dissolved palladium ions into the crystal structure or supported palladium on the perovskite surface. Both types of palladium states were found in the perovskite BaTi-Pd (see the XPS results in section 5.1.3). The Pd-supported perovskite Pd-BaTi clearly displayed a higher surface concentration of palladium than the Pd-substituted catalyst BaTi-Pd. These findings suggest the presence of a higher concentration of bulk-palladium in the perovskite BaTi-Pd than in Pd-BaTi.

6.3 The lanthanum perovskite coatings on top of EB-PVD PYSZ substrates

6.3.1 Coating methods of the lanthanum based perovskite $LaFeCo_{(0.3)}$-Pd

The lanthanum based perovskite of composition $LaFeCo_{(0.3)}$-Pd was coated on top of EB-PVD PYSZ substrates. Three different methods were employed for the coating development. For the first coating method the microstructure of the EB-PVD PYSZ ceramic substrates coated with the partially hydrolyzed gel displayed isolated perovskite islands. Disadvantage of this method is the low catalyst loading obtained after the drying and calcination of the coating. Even after multiple coating times an extremely low catalyst loading was obtained (1 mg of perovskite catalyst.g^{-1} PYSZ coating). A higher catalyst loading would be desirable in order to achieve high NO_x-conversions. The SEM-analysis of the coated EB-PVD PYSZ ceramic substrates applying this method showed a lot of catalyst-free areas (see Fig. 5.37) indicating the need for an optimum distribution of the perovskite catalyst. The uncoated EB-PVD PYSZ substrate should be used to increase the catalyst loading and thus the NO_x-conversions. The second coating method was carried out in order to improve the perovskite loading and distribution on the EB-PVD PYSZ coating surface.

The second coating method carried out by employing the previously synthesized perovskite powders provided higher catalyst loadings per gram of the EB-PVD PYSZ than the coating method applying solely the gel perovskite. The yield was approx. 2 mg of perovskite per gram of the PYSZ coating. Despite the achievement of higher catalyst loading an extremely poor adhesion of the perovskite particles onto the EB-PVD PYSZ coating surface was observed. The perovskite particles could be easily removed from the coating surface, although this method allowed the insertion of the perovskite particles between the gaps of the columnar structure of the EB-PVD PYSZ coating as documented in Fig. 5.38. Nevertheless, despite the fine particle size coating with the partially hydrolyzed perovskite gel was not sufficient to achieve this effect. It is likely that, the small particle sizes of the powder perovskite help to improve the distribution of the perovskite on the surface of the EB-PVD PYSZ coatings. The fillings of the gaps with the gel which crystallizes on further thermal treatment did not provide an adequate amount of material to fully cover the gaps between columns of the EB-PVD PYSZ coatings. Nevertheless, in spite of the fine particle sizes, the adhesion of the perovskite particles on the EB-PVD PYSZ coating was not strong enough. The investigations of the micrographs revealed that the necessary contact points between the particles were missing. In order to improve this aspect, a third coating approach combining the advantages of the first and second method was applied. In the third approach, fine perovskite particles were combined with the adhesion of the partially hydrolyzed gel substance. This method provided the formation of contact points between the perovskite powder particles and the substrate surface. Moreover, this method yielded the best quality in terms of the adhesion and homogeneity compared with those achieved by applying the first two methods.

The third coating method applied by using the previously synthesized perovskite particles + the partially hydrolyzed gel of the same composition provided final coatings of approx. 2.8 mg of perovskite per gram of the PYSZ substrate. A homogeneous catalyst coating was obtained applying this method. The perovskite based coating developed with this method was the best catalyst coating prepared in this work. The catalyst prepared applying this coating method provided the highest mass of the perovskite per gram of EB-PVD PYSZ substrate combined with the best quality i.e. homogeneity as compared to the coating applying the methods described before.

The XRD analysis showed that the perovskite coatings prepared with the third method were orthorhombic after heat treatment in air at 900°C/3h (see Fig. 5.37 in section 5.2). Moreover the XRD patterns included the tetragonal PYSZ phase with the typical preferred crystallographic growth in the <110> direction. The preferred orientation and crystal growth of the EB-PVD PYSZ columns depends mainly on the incidence angle of the vapour phase with respect to the surface sample. This effect has been already discussed and reported in the literature [95, 113]. No other phases belonging to single oxides (Fe, Co, La, etc) were detected by the XRD analysis, indicating the achievement of a coating with a stable perovskite phase on the EB-PVD PYSZ coating substrates.

6.3.2 Effect of the thermal ageing on the microstructure of $LaFeCo_{(0.3)}$-Pd-EB-PVD PYSZ double coatings

The thermal ageing of the synthesized $LaFeCo_{(0.3)}$-Pd-EB-PVD PYSZ double coatings were carried out with cycles combined of half an hour heating and 10 minutes cooling steps at 500°C and 900°C. At each temperature, the double coatings were tested with 25, 50 and 100 cycles. The main idea of these experiments was to prove the thermal shock resistance of the perovskite catalyst coated on top of EB-PVD PYSZ substrates. Under practical engine operation conditions the automotive catalyst may face environment conditions ranging from room temperature up to 900°C [9] (see the page 176). The synthesized coatings $LaFeCo_{(0.3)}$-Pd-EB-PVD PYSZ were thermally aged at 500°C and 900°C for half an hour and then cooled down to room temperature within 10 min. This thermal ageing procedure was repeated 25, 50 and 100 times at 500°C and 900°C. After the thermal ageing at 500°C up to 100 cycles no specific changes were observed in the coatings microstructure via the SEM analysis (see Figs. 5.40 to 5.42). No weight lost or gain of the coatings was registered after the thermal ageing at 500°C. The XRD analysis showed that as the number of cycles increases from 25 to 100 at 500°C the intensities of the diffraction patterns from the perovskite and the EB-PVD PYSZ substrates were increased. At a first glance, this may suggest the occurrence of a change (i.e. a phase transformation, a crystal stress release, and/or crystal growth, etc). However, since the ratio of the diffraction peaks (110/220) of the PYSZ textured columns remained almost constant, it can be postulated that no phase transformation is likely after ageing at 500°C (see Table 6.7). This agrees well with the weight of the coatings which remain constant after thermal ageing indicating that no oxidation processes or loss of the perovskite catalyst occurred.

Table 6.7 Intensity (in cps) of the diffraction signals of tetragonal PYSZ after cyclic thermal ageing at 500°C, 25, 50 and 100 cycles.

hlk	2θ	Thermal ageing at 500°C		
		25 cycles	50 cycles	100 cycles
110	35.14°	21045	27085	42604
220	74.2°	10811	13036	21473
(110/220)		1.9466	2.0777	1.9841

The orthorhombic crystal structure was stable after thermal ageing at 500°C. No changes in the peak ratio of the hkl 331/404 were observed, and the microstructure does not show any visible modifications after ageing (see Table 6.8). These observations are in agreement with the SEM observations. After thermal ageing at 500°C, no microstructural changes of the perovskite coated on top of the EB-PVD PYSZ coating were found by means of SEM.

Table 6.8 Intensity (in cps) of the diffraction signals of the orthorhombic perovskite coated on top of EB-PVD PYSZ after cyclic thermal ageing at 500°C, 25, 50 and 100 cycles.

hlk	2θ	Thermal ageing at 500°C		
		25 cycles	50 cycles	100 cycles
331	32.5°	827	827	1051
404	46.6°	402	361	453
(331/404)		2.0572	2.2908	2.3200

After the cyclic thermal ageing at 900°C, the ratio of the diffraction signals (110/220) was maintained constant over the cycles (see Table 6.9). The constant ratio suggests that no phase transformation occurs at 900°C. Even after 100 cycles of thermal ageing at 900°C of the coating LaFeCo$_{(0.3)}$-Pd-EB-PVD PYSZ, the hkl peak ratio of PYSZ (110/220) remain almost constant suggesting that no phase transformation occurs.

It is well known that a rapid reduction of the surface area occurs in the EB-PVD PYSZ coatings due to the porosity closure between 900°C and 1100°C (see Fig. 6.12) [114, 115]. It is previously reported that the BET surface area of the EB-PVD PYSZ coating decreases above 900°C yielding a substantial change in the morphology and porosity distribution. The most significant surface area change occurs at about 1100°C. Above 1100°C, the rate of the surface area reduction decreases significantly. The sintering behaviour and porosity modifications of the EB-PVD PYSZ coatings synthesized in the present study were not analyzed. The metastable t'-ZrO$_2$ phase of the EB-PVD PYSZ coating survives up to 1200°C [116] and that is why no crystal phase transformation of the EB PVD PYSZ coating are expected from room temperature up to 900°C.

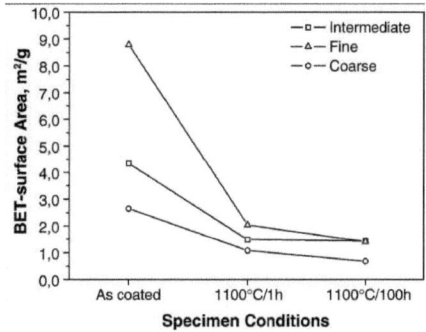

Figure 6.12 BET surface areas of EB-PVD PYSZ substrates with different microstructures as coated and after ageing on static air at 1100°C 1 h and 100 h, courtesy of [115].

The perovskite coating was only slightly affected with the cyclic thermal ageing at 900°C. Since no other phase(s) were detected by means of the XRD analysis, it can be assumed that the changes of the ratio (331/404) are mainly associated with the grain increase during crystallization of the perovskite coated on the EB-PVD PYSZ substrate (see Table 6.10). This agrees well with the SEM observations as mentioned in section 5 (see Figs. 5.44 to 5.46). Indeed, the microstructural analysis via the SEM displays the effect of the number of ageing cycles at 900°C on the morphology of the perovskite coated on the EB-PVD PYSZ substrates. The perovskite grains supported on top of the textured PYSZ columns started to migrate and grow.

Table 6.9 Intensity (in cps) of the diffraction signals of tetragonal PYSZ after cyclic thermal ageing at 900°C, 25, 50 and 100 cycles.

hlk	2θ	Thermal ageing at 900°C		
		25 cycles	50 cycles	100 cycles
110	35.14°	40512	109039	25117
220	74.2°	18002	48105	10624
(110/220)		2.2504	2.2666	2.3641

Table 6.10 Intensity (in cps) of the diffraction signals of the orthorhombic perovskite coated on top of EB-PVD PYSZ after cyclic thermal ageing at 900°C, 25, 50 and 100 cycles.

hlk	2θ	Thermal ageing at 900°C		
		25 cycles	50 cycles	100 cycles
331	32.5°	1536	1538	1021
404	46.6°	629	669	590
(331/404)		2.4419	2.2989	1.7305

6.3.3 NO$_x$-reduction performance of the powder catalyst vs. the catalytic coatings

The catalytic tests for NO$_x$-reduction given in previous sections were carried out using powder samples. The reason was that with a relatively low mass of catalyst, reasonably good results can be obtained. It is also possible to carry out many tests by spending less gas emissions and reductants enabling an economic and fast testing of a numerous catalyst compositions. In this way, time and resources can be saved. In addition, if the mass of the catalyst is low enough some other effects such as mass and heat transfer that may interfere during the reaction can be neglected to some extend. Therefore the plug flow reactor is the preferred choice for the kinetic studies. Nevertheless it is necessary to gain an idea about the performance of the catalyst near real reaction conditions. Particularly, catalytic coatings for automotive applications must work under completely different hydrodynamic conditions in comparison to the plug-flow reactor, which is usually the case for powder testing.

In the present work, both the perovskite powder catalyst LaFeCo$_{(0.3)}$-Pd calcined in air at 700°C for 3h, and the cordierite monolith coated with the same perovskite composition calcined in air at 700°C for 3h, were tested for the C_3H_6-SCR of NO$_x$ reaction. The NO$_x$-reduction began between 200°C and 225°C with the powder catalyst LaFeCo$_{(0.3)}$-Pd for a reaction mixture containing 250 ppm NO, 500 ppm C_3H_6, 1 vol. % O_2 and Ar balance (Flow rate = 300 ml.min^{-1}). The maximum NO$_x$-conversion of 34 % was reached at 250°C with the powder catalyst under the reaction conditions mentioned above. This reaction conditions lead to a space velocity of 300 000 h^{-1} considering the volume occupied for 75 mg of the pelletized powder catalyst inside the quartz tube reactor.

On the contrary, the NO$_x$-reduction began at 260°C with the perovskite LaFeCo$_{(0.3)}$-Pd coated on top of the cordierite monolith (see Fig. 5.70). The maximum NO$_x$-conversion of 20 % was reached between 350°C and 450°C under a feed composition of 510 ppm NO, 520 ppm C_3H_6, 1.1 vol. % O_2, 3.9 vol. % H_2O_{vapour}, and N_2 as balance at a space velocity of 60000 h^{-1}. The increase in the oxygen concentration to 5 vol. % in the feed shifted of the maximum NO$_x$-conversion to 300°C on the powder catalyst LaFeCo$_{(0.3)}$-Pd. Above 300°C the NO$_x$-conversion decreased up to 10 % on the perovskitic powder catalyst LaFeCo$_{(0.3)}$-Pd, while the coated monolith displayed the beginning of NO$_x$-conversion at 250°C reaching the maximum of 15 % between 400°C and 450°C. The NO$_x$-reduction performance of the perovskitic powder catalyst is evidently different from the NO$_x$-performance of the catalytic perovskite coated on the cordierite monolith.

Most of the new catalyst development studies are carried out by employing powdered samples. There is a limited number of publications dealing with the model type of the perovskite based catalysts coated on monoliths (model catalytic converters) [13, 72]. The present literature focuses on the determination of the catalytic activity towards CO/NO$_x$ reactions for TWC applications. This implies the use of rich feed compositions with sometimes extremely low partial pressures of oxygen (less than 1 vol. % O_2). Although the development trend directs the lean-burn applications, there are very few published studies dealing with emission reduction under lean conditions [117]. Furthermore, there is extremely large variety of reaction conditions used in the NO$_x$-conversions tests of catalytic materials. Table 6.13 shows typical reaction conditions (i.e. gas flow, precious metal loading) and the characteristics of the tested monolith in each case. In general, in order to provide a comparison base, a unit called space velocity is applied. The space velocity combines the volume of gas used during the catalytic test is coupled to the volume of the catalyst. A space velocity of 5 h^{-1} means that five reactor volumes of feed at specified conditions are being fed into the reactor per hour.

The thesis of Frank comprehensively deals with a C_3H_6-SCR of NO$_x$ study employing the Pt-supported catalyst coated on a cordierite monolith [117]. The test conditions used in this study differ from those of this work. The reaction conditions are compared in Table 6.13. As using 5 vol.

% O_2 in the feed, Frank obtained 65 % of NO_x-conversion at 270°C over the Pt-supported catalyst. The NO_x-conversion started as early as 175°C under the feed conditions of (500 ppm NO, 500 ppm C_3H_6, 5 vol. % and N_2 as balance). The reaction products over the platinum based catalyst were N_2, N_2O and NO_2. A maximum selectivity of 50 % to N_2O-formation was measured at 240°C. This is the main problem of platinum catalysts which produce a lot of N_2O under rich and lean conditions.

In comparison, the Pd-supported/substituted perovskite-based catalysts coated on cordierite substrate yielded a NO_x-conversion of 20 % at the temperature range of 350°C and 450°C. This means that only 1/3 of the conversion achieved over the Pt-based catalyst given in Frank's study was reached [117]. The enormous difference can be partially due to the reaction conditions, since Frank did the catalytic test with a 9.42 ml of the catalytic converter at a flow rate of 6 l per minute thus providing a lower space velocity (~38216 h^{-1}). In the present case the catalytic converter (perovskite coated on the cordierite monolith) was tested for the C_3H_6-SCR of NO_x reaction at a space velocity = 60000 h^{-1}, this is 1.57 times higher than the space velocity employed by Frank [117]. In general, the platinum supported catalysts are more active for C_3H_6-SCR of NO_x applications and less selective to nitrogen than palladium based catalysts [1].

Table 6.13 Monolith catalysts characteristics and reaction conditions for different reactions

Catalyst Volume (cm^3)	Cell density of the monolith (cell/in^2)	Metal loading (g/ft^3)	Reaction	Space Velocity (h^{-1})	Ref
567	400	90 g Pd [a]	CO/NO	$2.1.10^5$ h^{-1}	[72]
12.75	400	31.15 Pd [b]	CO/NO	50000 h^{-1}	[13]
9.42	-----	Pt [c]	C_3H_6-SCR of NO_x	38216 h^{-1}	[117]
36.6	400	70 g Pd	C_3H_6-SCR of NO_x	60000 h^{-1}	This work

[a] Perovskite catalyst containing Pd supported on γ-Al_2O_3
[b] Catalyst with Al_2O_3 wash-coat
[c] Catalyst with Al_2O_3 wash-coat

Chapter 7

Conclusions

Material synthesis

A series of lanthanum and barium based perovskite catalysts are synthesized by a modified citrate route and co-precipitation method respectively. These perovskitic structures are prepared for the first time at the facilities of the high temperature and functional coatings division of DLR's Institute of Materials Research. A modified citrate route is employed to the synthesis of the lanthanum based perovskites. Care must be taken when working with nitrates in the presence of cobalt and iron since explosive mixtures can be formed. An additional care must be taken for the synthesis of the barium based perovskites. The co-precipitation method employed to synthesize these barium perovskites implies the use of metallic barium pieces. To avoid any exothermic reaction(s) and minimize the risk of explosion the metallic barium must not be brought into contact with aqueous solutions or water.

A single La-based perovskite phase with orthorhombic structure is synthesized by applying the citrate route. The Lanthanum based perovskite $LaFeCo_{(0.3)}$-Pd is amorphous upon calcination in air at 500°C and begins to crystallize into the orthorhombic structure at 550°C. The crystallization process of the perovskite $LaFeCo_{(0.3)}$-Pd ends at approx. 700°C. No other oxide phase(s), i.e. palladium oxide, are found after calcination in air up to 700°C by the XRD analysis, suggesting the formation of a solid solution between palladium and the perovskite. At higher calcination temperatures tetragonal PdO particles begin to diffuse out of the perovskite lattice and to grow, indicating the meta-stability of the produced perovskitic solid solution ($LaFeCo_{(0.3)}$-Pd).

The crystal sizes of the Ba-based perovskites $BaTiO_3$ obtained applying the co-precipitation method are remarkably smaller than the crystal sizes of the La-based perovskites obtained via the citrate route. The small crystal sizes of the Ba-based perovskites leads to a crystallization path that begins already at 500°C. Fully crystallized Ba-based perovskites are obtained, whereas the La-based perovskites are amorphous after calcination in air at 500°C. The perovskite phase $BaTiO_3$ is composed of a pure tetragonal perovskite with traces of $BaCO_3$. No other oxides such as BaO, TiO_2 or other mixed oxides are observed upon calcination up to 900°C, pointing out the success of this synthesis route to prepare the single phase perovskite. Barium carbonate is completely removed from the catalyst upon calcination in air at 900°C/3h.

Self-healing of palladium in the La- and Ba-based perovskites

The Pd-substituted perovskites are known with their adaptive character which induces an ageing resistance of the catalyst material. The occurrence of this characteristic is investigated in this study. According to the SEM and XRD observations the Pd-substituted La-based perovskite $LaFeCo_{(0.3)}$-Pd-700°C displays the diffusion of palladium out of the crystal lattice upon reduction treatment. Upon re-oxidation this second phase disappears indicating that the palladium ions may reversibly diffuse inside of the perovskitic crystal lattice (the self-healing property). For comparison purposes, the Pd-impregnated perovskite powders are synthesized to yield the palladium supported perovskite Pd-LaFeCo$_{(0.35)}$. In contrast to the Pd-substituted perovskite, the as synthesized Pd-supported catalyst displays a second phase at the grain boundaries and on the surface of the perovskite indicating the presence of palladium particles sizes up to 40 nm.

The XPS analysis of the as prepared Pd-substituted perovskite LaFeCo$_{(0.3)}$-Pd indicates that palladium ions are partially incorporated into the crystal structure in this perovskite. On the other hand, in the Pd-supported perovskite, palladium is mainly on the surface and at the grain boundaries, however a partial dissolution of palladium has been observed on the first lattice layers of the perovskite particles. Palladium is reduced to metallic Pd° at the supported perovskite Pd-LaFeCo$_{(0.35)}$ under a hydrogen containing atmosphere already at 200°C confirming the presence of mostly PdO on the catalyst surface in the oxidized state. In contrast, the palladium species in the Pd-substituted perovskite LaFeCo$_{(0.3)}$-Pd survive even after reduction in hydrogen at 500°C. Palladium is partially dissolved into the crystal lattice in the Pd-supported perovskite Pd-LaFeCo$_{(0.35)}$ upon calcination treatments. Moreover, different concentrations of palladium are recorded at the surfaces of both perovskites. The relative palladium concentration at the perovskite surface in the catalyst LaFeCo$_{(0.3)}$-Pd is considerably lower than the palladium concentration in the catalyst Pd-LaFeCo$_{(0.35)}$ as determined by XPS indicating that a higher concentration of palladium is present in the bulk of the former than in the latter. Despite the fact that less Pd is readily available on the perovskite surfaces, the catalytic activity of the Pd-substituted perovskite [LaFeCo$_{(0.3)}$-Pd] is not inferior to that of Pd-supported perovskite [Pd-LaFeCo$_{(0.35)}$]. On the contrary, the Pd-substituted perovskite catalyst displays higher NO-conversions than the Pd-supported perovskite towards C_3H_6-SCR of NO with 5 vol. % O_2.

In the Pd-supported perovskite Pd-BaTi the XPS line of Pd $3d_{5/2}$ at 337.8 eV suggests that palladium is partially dissolved in the first layers of the perovskite lattice upon calcination treatment. These palladium species disappear upon reduction already at 200°C forming metallic palladium Pd° (BE Pd $3d_{5/2}$ = 334.2 eV). The XPS analysis of the Pd-substituted catalyst BaTi-Pd suggests that palladium ions are not fully incorporated into the perovskite lattice but highly dispersed on the perovskite surface. The relative concentration of palladium at the surface of the Pd-substituted catalyst (BaTi-Pd) compared to that in the Pd-supported catalyst (Pd-BaTi), also indicates that most of palladium in the Pd-integrated perovskite (BaTi-Pd) is inside the bulk structure. Even though less Pd is available on the catalyst surface no detrimental effect on the C_3H_6-SCR of NO_x reaction is observed with the Pd-substituted perovskite. Pd in the lattice and on the surface of both Ba-based perovskites is reduced to its metallic state Pd° after the reduction treatments.

Catalytic activity of the lanthanum based perovskites

The Co-content in the La-based perovskites has a significant influence on their catalytic properties. There is a solubility limit of cobalt in the perovskite by the substitution at the B-site at the expenses of Fe. Cobalt oxide (Co_3O_4) has been detected in compositions with the Co-substitutions above 0.3 mol % in the B-site of the La-based perovskite. The decrease in Co-content from 0.475 to 0.3 fraction mol in the perovskite resulted in 14 % less NO_x-conversion at about 200°C. Although, the catalyst powders contains only a single perovskite phase as the Co-content decreased down to the 0.3 mol, as mentioned above, the catalytic activity towards the NO_x-reduction suffer slightly. This indicates that the perovskite composition actively participates in the

NO$_x$-conversion process. The phase(s) composition of the La-based perovskite changes drastically with modification of the Co-content. The perovskite with Co-content of 0.475 mol % LaFeCo$_{(0.475)}$-Pd is composed of the mixed orthorhombic and cubic perovskites. The split of the main XRD reflections, only observed upon heating, indicates the occurrence of both perovskites. This condition (i.e. Co$_3$O$_4$ + the orthorhombic and cubic perovskites) in the catalyst holds in the presence of water vapour and CO$_2$ in the feed resulting in a better NO$_x$-reduction and higher N$_2$-selectivity of the catalyst LaFeCo$_{(0.475)}$-Pd compared to the perovskite with lower Co-content. Although at temperatures above 240°C the presence of water vapour and CO$_2$ has as matter of fact a detrimental influence on the NO$_x$-conversion capability of the low Co-content La-based perovskite catalyst (LaFeCo$_{(0.3)}$-Pd) below 240°C. No measurable influence of the H$_2$O + CO$_2$ addition on the NO$_x$-conversion is detected. Positive effects of water vapour on the conversion of NO could be attributed to the presence of unknown metallic (or maybe bimetallic) species which dissociate the water molecules resulting in an increase of the coverage of chemisorbed hydrogen atoms. The hydrogen coverage may promote the dissociation of NO which is considered the rate determining step in the reaction. Identification and characterisation (i.e. conversion, selectivity, and adsorption-desorption properties measurements determination, etc) of the mentioned metallic species may provide answers to these questions.

Addition of CO in the feed (NO/H$_2$) results in a decrease in the catalytic NO$_x$-reduction and N$_2$-selectivity of both Co-containing LaFeCo$_{(0.475)}$-Pd and LaFeCo$_{(0.3)}$-Pd La-based perovskite catalysts. Higher Co-content in the perovskite yields a better resistance against CO. The improved catalytic activity of the Pd-substituted perovskite with 0.475 mol % Co-content can be associated to some degree with the segregated Co$_3$O$_4$. A positive effect on the NO$_x$-conversion due to phase cooperation may also be argued, however systematic studies are needed to elucidate this assumption. The LaFeCo$_{(0.475)}$-Pd catalyst reduces 8 to 13 % more NO$_x$ and produces 10 to 50 % more N$_2$ compared to the performance of the perovskite with the lower cobalt content LaFeCo$_{(0.3)}$-Pd.

The cerium solubility in the A-site of the La-based perovskite is very limited to less than 0.05 mol % Ce. In the present study a single phase is obtained in the perovskites La$_{(1-x)}$Ce$_x$Fe-Pd (x = 0.05, 0.1 and 0.4), after calcination in air up to 700°C. However, calcination of these perovskite compositions at 900°C showed that these phases are meta-stable and some traces of CeO$_2$ and Fe$_2$O$_3$ segregate leaving back a LaFe-rich perovskite. A higher Ce-concentration (up to 0.4 mol %) [e. g. LaCe$_{(0.4)}$Fe-Pd] delays the crystallization of the perovskite. As a matter of fact, the catalyst is amorphous displaying a disordered crystallographic characteristic. The crystallization of the catalyst occurs at 900°C by yielding rhombohedral Fe$_2$O$_3$, cubic CeO$_2$ and an orthorhombic perovskite. This catalyst displays a higher BET-surface area (6.647 m^2/g) compared to the La-based perovskite with the lowest Ce-content [LaCe$_{(0.05)}$Fe-Pd] (1.598 m^2/g) on calcination at 900°C. The reason for this behaviour is the earlier crystallization temperature of low Ce-containing La-based perovskite which goes into a grain growth on calcination at 900°C. The best NO$_x$-reduction performance and N$_2$-selectivity for H$_2$-SCR of NO$_x$ reactions may be relying on the relative high surface area of the catalyst LaCe$_{(0.4)}$Fe-Pd. The catalyst LaCe$_{(0.4)}$Fe-Pd displays high contents of Fe- and Ce-oxides which strongly suggest the participation of these oxides in the improved NO$_x$-reduction performance of the catalyst. It would be mere speculation to say that these oxides may eliminate

N_2O under the selected reaction conditions. The N_2O-decomposition determination of the single and mixed oxides of Fe- and Ce- under lean conditions could give insights in this direction.

Both the Pd-supported and Pd-substituted La-based perovskites yield similar NO_x-conversion performances under C_3H_6-SCR conditions and 1 vol. % O_2 in the feed. A maximum NO_x-conversion of 35 % is measured over these two perovskites at 250°C. Although the catalyst Pd-LaFeCo$_{(0.35)}$ displayed 20 % more NO-conversion than the palladium integrated catalyst LaFeCo$_{(0.3)}$-Pd with 1 vol. % O_2 at 225°C. An improved NO_x-reduction capability at higher oxygen concentrations (2 and 5 vol. % O_2) is observed over the Pd-substituted perovskite LaFeCo$_{(0.3)}$-Pd than over the Pd-supported catalyst Pd-LaFeCo$_{(0.35)}$. A maximum NO_x-conversion of 28 % of NO_x-conversion is detected at 225°C with 2 vol. % O_2 in the feed over the catalyst LaFeCo$_{(0.3)}$-Pd, yielding 2 to 7 % more NO_x over the whole temperature range between 175° and 300°C than the supported catalyst Pd-LaFeCo$_{(0.35)}$. The catalyst Pd-LaFeCo$_{(0.35)}$ reduced 5 % more NO_x than the palladium substituted perovskite LaFeCo$_{(0.3)}$-Pd with 2 vol. % O_2 in the feed only at 200°C. No explanation can be given to date for this result. Systematic and more experiments under similar reaction conditions are needed to convincingly validate these observations. With 5 vol. % oxygen in the gas mixture the maximum NO_x-conversion of 26 % is registered over the catalyst LaFeCo$_{(0.3)}$-Pd at 300°C, this perovskite reduces 2 to 10 % more NO_x over the whole temperature range between 175°C and 300°C than the supported catalyst Pd-LaFeCo$_{(0.35)}$ with 5 vol. % O_2, the highest oxygen concentration tested in this study.

Catalytic activity of the barium based perovskites

The Pd-substituted Ba-based perovskite displayed 45 to 50 % of NO_x-conversions between 200° and 250°C under simple gas mixture (e. g. simple gas mixture includes NO, O_2 and H_2 with He as balance). A maximum NO_x-conversion of 73.2 % and N_2-selectivity of 82 % is measured over the Pd-substituted catalyst BaTi-Pd during the wet NO_x-reduction at 171°C. The presence of $H_2O + CO_2$ causes a positive effect on the NO_x-conversion and on the selectivity to nitrogen over this catalyst between 200°C and 270°C. At least in the presence of water the N_2O molecules are formed and not participate in any other reactions over the Ba-based catalyst. Positive effects on the NO_x-conversion over the Ba-based catalyst are probably related to the NO dissociation enhancement caused by water molecules. These assumptions need to be clarified in further studies.

The presence of CO in the reaction mixture causes a strong decrease in the NO_x-reduction and selectivity to nitrogen of this catalyst below 200°C, suggesting selective chemisorption of CO on probably the same catalytic active sites. The N_2O-formation obtained in the presence of CO is similar to the reaction in the presence of water + CO_2 (approx. 60 ppm), indicating that CO does not participate in the N_2O-formation over this catalyst. At a higher temperature the NO_x-conversion is negatively affected with CO in the feed and in comparison with some of the La-perovskites, less N_2O is produced which lead to a better selectivity to nitrogen with the Ba-based perovskite. This catalyst reduced 10 to 20 % more NO_x in the absence of CO in the temperature range between 150°C and 350°C.

In general the perovskite BaTi-Pd is one of the best catalysts in terms of NO_x-reduction and N_2-selectivity under the selected reaction conditions for H_2-SCR of NO_x reactions. Under simple reaction conditions (NO/O_2/H_2/He) the catalyst BaTi-Pd displays similar NO_x-reduction levels (80 %) compared to the NO_x-performance over the perovskite LaFeCo$_{(0.3)}$-Pd. On these reaction conditions the Pd-substituted perovskite BaTi-Pd produces 22 % more nitrogen (or less N_2O) than the Pd-substituted La-based catalyst [LaFeCo$_{(0.3)}$-Pd]. The BaTi-Pd catalyst heat treated at 500°C/3h shows 25 to 30 % higher NO_x-conversion at catalyst bed temperatures between 175° and 220°C than the same catalyst composition treated at 900°C/3h.

The Barium group of perovskites displays also the maximum NO_x-conversion during the C_3H_6-SCR of NO_x reaction between 200°C and 225°C with 1-5 vol. % O_2. The propene light-off temperatures correspond well with the maximum NO_x-conversion values obtained with the perovskites. The Pd-substituted Ba-based perovskite BaTi-Pd displays the highest NO_x-conversion (42.4 %), with 1 vol. % O_2 in the feed, of all the tested perovskites in this work. Increase of the oxygen concentration to 2 and 5 vol. % increases the oxidation potential of the catalyst, which causes indeed the oxidation of propene at lower temperatures followed by a decrease in the NO_x-conversion. The Pd-substituted perovskite BaTi-Pd displays a better NO_x-conversion than the perovskite Pd-BaTi with 1 and 5 vol. % O_2 in the feed. The catalyst BaTi-Pd reduces 4 to 9 % more NO_x than the perovskite Pd-BaTi during the reaction with 1 vol. % O_2 in the gas mixture. With 5 vol. % O_2 in the feed the catalyst BaTi-Pd reduces 4 to 5 % more NO_x between 150°C and 200°C than the Pd-supported perovskite Pd-BaTi. At 225°C and higher temperatures the NO_x-conversions are almost the same for both catalysts with 5 vol. % O_2 in the feed. The NO_x-conversions obtained over these perovskites are similar with 2 vol. % O_2 in the gas mixture. These observations are unexpected because it implies that both catalysts pose the same NO_x-conversion potential. On the other hand, the Pd-substituted catalyst BaTi-Pd displays a slightly better NO-conversion at 200°C and lower reaction temperatures than the Pd-supported catalyst Pd-BaTi demonstrating the potential of the former.

The catalytic coating

Three methods are employed to coat the EB-PVD PYSZ ceramic substrates with the La-based perovskite catalyst; the first method includes the previous synthesis of the perovskite applying the citrate route described in this study. Coating of the EB-PVD PYSZ ceramic substrates with the partially hydrolyzed gel of composition LaFeCo$_{(0.3)}$-Pd leads to the formation of isolated perovskite agglomerates supported mainly on top of the EB-PVD PYSZ columns. The gaps between the EB-PVD PYSZ columns are not filled by applying this coating method leading to a very low perovskite loading on the surface of the PYSZ substrate.

The second coating method is carried out employing only the suspension containing La-based perovskite particles previously synthesized via the citrate route and pre-calcined at 500°C. This approach provides a coating with inhomogeneous thicknesses where the perovskite particles are not really added to the PYSZ surface. The perovskite particles are deposited between the gaps of the EB-PVD PYSZ substrate whereas the column tips remain uncoated. This is the opposite to that microstructure obtained with the first coating method where most of the column tips are covered with the gel of the La-based perovskite.

To improve the quality of the La-based perovskite coatings on the EB-PVD PYSZ substrates, a third approach is carried out combining the observed advantages of the first two coating methods applied in this study. It combines the adhesion of the partially hydrolyzed perovskite observed in the first coating approach and the higher distribution degree achieved with the La-based perovskite particles alone. The coating method combining the two approaches mentioned above is classified as the best method employed in this study. Partially hydrolyzed gel of the perovskite acted as a "glue" between the previously synthesized and pre-calcined perovskite particles and also between the perovskite particles and the EB-PVD PYSZ substrate. Additionally, these coatings display a very good resistance during cyclic thermal shock ageing at 500° and 900°C. No phase(s) and morphological changes are observed upon thermal ageing at 500°C. The perovskite grains coated on the EB-PVD PYSZ surface start to migrate and grow after thermal cycling at 900°C. However, no significant crystal phase(s) changes of the coatings are observed after thermal ageing at this temperature. No mass gain or loss is registered after thermal ageing of the coatings $LaFeCo_{(0.3)}$-Pd-EB-PVD PYSZ at 500° and 900°C demonstrating the stability of the perovskite crystals against cyclic thermal loading. The coatings $LaFeCo_{(0.3)}$-Pd-EB-PVD PYSZ are tested for C_3H_6-SCR of NO_x reduction but no activity is observed due to the low perovskite loading at the EB-PVD PYSZ substrates and to the low geometric surface area of the coatings. It is recommended to increase the geometric surface area for optimal catalytic testing of the coatings.

Catalytic coatings with higher geometrical surface than the coated EB-PVD PYSZ substrates are obtained by coating cordierite ceramic substrates with the La-based perovskite catalyst. The perovskite particles $LaFeCo_{(0.3)}$-Pd are successfully coated on top of cordierite substrates provided by INTERKAT©. Colloidal silica is employed as a binder between the powder particles and the ceramic substrate, this method provided catalyst coatings with thickness of a few micrometers up to 20 μm. The coated cordierite substrates are catalytically tested for C_3H_6-SCR of NO_x at high space velocities. In the presence of water vapour and high oxygen concentration (4.7 vol. %) a maximum 20 % of NO_x-conversion is measured at ~ 450°C with the new catalytic converter. The powder based catalysts (the La-based perovskite alone) show slightly higher NO_x-conversions at lower temperatures than the catalytic converter. In the catalytic converter the reactants must flow through the open channels and diffuse to the wall before they can react on the active sites. On the other hand the powder bed samples are supported between two plugs of quartz wool providing high contact surface between the catalyst and the gas phase resulting in improved NO_x-conversions rates.

Chapter 8

Outlook

This study shows that palladium is partially incorporated into the crystal lattice of the synthesized Pd-substituted perovskites. Additionally, palladium ions are also partially incorporated in the crystal lattice of the perovskites after the impregnation of palladium and calcination treatment. However the Pd-substituted perovskites display a better palladium distribution than the Pd-impregnated perovskite as observed in the SEM-analysis. These observations demonstrate the importance of the synthesis method to successfully incorporate palladium or other ions into perovskite lattices. In order to increase the concentration of palladium in the crystal lattice it is recommended to work with the chemistry of the precursors. The use of palladium acetate or acetyl-acetonates instead of nitrate(s) may lead to a better degree of ions substitution in the perovskite lattices. Mechanical alloying may be an alternative to force the palladium ions to get into the perovskite lattice. On the other hand, the optimization of the palladium loading (decrease) in the perovskite(s) measuring its effect on the NO_x-reduction and selectivity, and durability of the catalyst would be of importance.

On redox treatments palladium ions may diffuse in and out of the perovskite lattice(s). In this work the relative surface concentrations of Pd determined by the XPS analysis indicate that many other diffusion processes may occur upon reduction treatments. It still remains open the question if palladium forms nano-alloys (and which) upon reduction; La_3Pd, Pd-Co and Pd-Fe are just a few possible combinations. Further TEM-analysis of the reduced perovskites and the corresponding reference samples i.e. La_3Pd, Pd-Co or Pd-Fe may help to prove this hypothesis.

Further improvements in the NO_x-conversion and N_2-selectivities of the La- and Ba- based perovskites, as well as the development of new catalysts, are required in order to facilitate the possible commercialization of the H_2-SCR of NO_x technology in the coming years. The challenge will be to design a new generation of catalyst(s) that produces lower (or even better zero) concentrations of N_2O. High loadings of cerium in perovskite catalysts seem to be a promising combination to reduce the formation rates of N_2O during the H_2-SCR of NO_x reaction. So, the optimization of the ceria loading in the perovskite based catalysts would be the next step to further reduce or eliminate the N_2O-formation during H_2-SCR of NO_x. An alternative to solve this issue would be to design catalysts that are able to decompose N_2O under excess oxygen and in the presence of water vapour, CO_2, and CO, etc. Wider operating temperature windows of the catalytic converter would be helpful for an efficient control emission of NO_x under the different driving modes on the European roads. The combination of various catalysts compositions as those suggested in this study may yield a larger temperature window of application where the catalyst effectively reduces NO_x. Long-time stability tests of the perovskite catalysts against the thermal deactivation and the exposure to poisons such as SO_2 are also needed.

Further improvements in the NO_x-reduction and selectivity to nitrogen under lean conditions employing hydrocarbons or other reductants are needed. The use of methane as a reducing agent is an appealing alternative for the HC-SCR of NO_x-reaction. However, to date is quite difficult to activate the methane molecules, especially at low temperatures (i.e. the cold start). Moreover, it is necessary to determine the NO_2- and N_2O- selectivities of the perovskite based catalysts during the C_3H_6-SCR of NO_x reactions. Depending on the reaction conditions (i.e. gas concentration, flow and mass of catalyst) high volumes of CO can also be formed. One has to make sure that CO completely oxidizes under the selected reaction conditions. The use of propene and higher hydrocarbons as reducing agents for HC-SCR applications is still a promising approach to eliminate NO_x under lean conditions. Unburned hydrocarbons already present in the exhausts of combustion engines or intentionally added to it could be applied as reductants for NO_x-reduction. The main issue is to avoid the selective oxidation of the reductant(s) and promote the NO_x-reduction mainly at low and high operating temperatures. Catalysts with chemical elements that are less selective to the hydrocarbon oxidation and more active for the selective reduction of NO_x are needed.

The successful application of the catalytic coatings to eliminate NO_x under lean conditions depends on additional properties. For instance, the mechanical properties of the catalytic coatings are not studied in this work. Conventional catalytic converters are exposed to extremely high volumes of hot gases that cause stresses and erosion effects to the catalytic coatings. For example a cyclic thermal loading in the presence of water vapour would help to simulate these environments. Similarly, the resistance of these coatings against erosion under strong corrosive environments at moderate and high temperatures would be worth to study before the transfer of this catalyst technology in automotive and/or aeronautical applications.

In order to determine the feasibility of the NO_x-reduction in exhausts produced by airplanes with catalytic converters, a critical study of the advantages and disadvantages of the possible application of catalyst for NO_x-reduction in aircrafts is needed. Type of engine, fuel composition, power setting(s), all these parameters are directly related to the fuel consume in aircrafts and are factors that would influence the performance of an aeronautic catalytic converter. Moreover a catalyst in aeronautical applications may be exposed to various deactivation processes. Chemical deactivation caused by the presence of compounds that are already present in typical aviation fuels i.e. additives, lubricants, etc. Common additives to aviation fuel include alkyl-lead anti-knock additives, metal deactivators, oxidation inhibitors, corrosion inhibitors, icing inhibitors, biocides, and static dissipaters. Surely, the out coming gas emissions from commercial aircraft engines are a growing part in the transportation sector and must be reduced in the forthcoming future.

References:

1. Burch, R. and P.J. Millington, *Selective reduction of nitrogen oxides by hydrocarbons under lean-burn conditions using supported platinum group metal catalysts.* Catalysis Today, 1995 **26** (2): p. 185-206.
2. Burch, R., J.P. Breen, and F.C. Meunier, *A review of the selective reduction of NO_x with hydrocarbons under lean-burn conditions with non-zeolitic oxide and platinum group metal catalysts.* Applied Catalysis B: Environmental, 2002 **39** (4): p. 283-303.
3. Zhang, R., et al., *Cu- and Pd-substituted nanoscale Fe-based perovskites for selective catalytic reduction of NO by propene.* Journal of Catalysis, 2006 **237** (2): p. 368-380.
4. Burch, R. and M.D. Coleman, *An investigation of the $NO/H_2/O_2$ reaction on noble-metal catalysts at low temperatures under lean-burn conditions.* Applied Catalysis B: Environmental, 1999 **23** (2-3): p. 115-121.
5. Costa, C.N., et al., *An Investigation of the $NO/H_2/O_2$ (Lean-DeNO$_x$) Reaction on a Highly Active and Selective $Pt/La_{0.5}Ce_{0.5}MnO_3$ Catalyst.* Journal of Catalysis, 2001 **197** (2): p. 350-364.
6. Costa, C.N., et al., *An Investigation of the $NO/H_2/O_2$ (Lean De-NO$_x$) Reaction on a Highly Active and Selective $Pt/La_{0.7}Sr_{0.2}Ce_{0.1}FeO_3$ Catalyst at Low Temperatures.* Journal of Catalysis, 2002 **209** (2): p. 456-471.
7. Perez-Alonso, F.J., et al., *Synergy of $Fe_xCe_{1-x}O_2$ mixed oxides for N_2O decomposition.* Journal of Catalysis, 2006 **239** (2): p. 340-346.
8. Yan, L., et al., *Catalytic decomposition of N_2O over $M_xCo_{1-x}Co_2O_4$ (M = Ni, Mg) spinel oxides.* Applied Catalysis B: Environmental, 2003 **45** (2): p. 85-90.
9. Bode, H., *Materials Aspects in Automotive Catalytic Converters*, ed. DGM. Vol. 1. 2001: Wiley-VCH. 278.
10. Yokoi, Y. and H. Uchida, *Catalytic activity of perovskite-type oxide catalysts for direct decomposition of NO: Correlation between cluster model calculations and temperature-programmed desorption experiments.* Catalysis Today, 1998 **42** (1-2): p. 167-174.
11. Tanaka, H., Uenishi, M., Kajita, N., Tan, I., Nishihata, Y., Mizuki, J., Narita, K., Kimura, M., Kaneko, K., *Self-Regenerating Rh- and Pt-Based Perovskite Catalysts for Automotive-Emissions Control.* Angew. Chem. Int. Ed., 2006 **45**: p. 5998-6002.
12. Nishihata, Y., et al., *Self-regeneration of a Pd-perovskite catalyst for automotive emissions control.* Nature, 2002 **418** (6894): p. 164-167.
13. Tagliaferri, S., R. Koppel, and A. Baiker, *Influence of rhodium- and ceria-promotion of automotive palladium catalyst on its catalytic behaviour under steady-state and dynamic operation.* Applied Catalysis B: Environmental, 1998 **15** (3-4): p. 159-177.
14. Bosch, H. and F. Jansen, *Formation and control of nitrogen oxides.* Catalysis Today, 1988 **2** (4): p. 369-379.
15. Kapteijn, F., J. Rodriguez-Mirasol, and J.A. Moulijn, *Heterogeneous catalytic decomposition of nitrous oxide.* Applied Catalysis B: Environmental, 1996 **9** (1-4): p. 25-64.
16. Norms, E., *Draft (Commission Regulation EC No...).* 2008: Brussels. p. 164.
17. ICAO, *ICAO Council adopts New Standards for Aircraft Emissions.* 2005.
18. ICAO, *ICAO Engine exhaust emissions data bank, subsonic engines.* 2008.
19. Frenkel, M. 1909: France, England.
20. Shelef, M. and R.W. McCabe, *Twenty-five years after introduction of automotive catalysts: what next?* Catalysis Today, 2000 **62** (1): p. 35-50.
21. Wulff, A. and J. Hourmouziadis, *Technology review of aero-engine pollutant emissions.* Aerospace Science and Technology, 1997 **1** (8): p. 557-572.
22. Miura, N., M. Nakatou, and S. Zhuiykov, *Impedancemetric gas sensor based on zirconia solid electrolyte and oxide sensing electrode for detecting total NO_x at high temperature.* Sensors and Actuators B: Chemical, 2003 **93** (1-3): p. 221-228.

23. Taylor, K.C., *Nitric Oxide Catalysis in Automotive Exhaust Systems.* Catalysis Reviews, 1993 **35** (4): p. 457 - 481.
24. Ueda, A., et al., *Two conversion maxima at 373 and 573 K in the reduction of nitrogen monoxide with hydrogen over Pd/TiO$_2$ catalyst.* Catalysis Today, 1998 **45** (1-4): p. 135-138.
25. Heck, R.M., *Catalytic abatement of nitrogen oxides-stationary applications.* Catalysis Today, 1999 **53** (4): p. 519-523.
26. Penner, J.E., et al., *Aviation and the global atmosphere.* 1999, Intergovernmental Panel on Climate Change. p. 373.
27. Busca, G., et al., *Catalytic abatement of NO$_x$: Chemical and mechanistic aspects.* Catalysis Today, 2005 **107-108**: p. 139-148.
28. Gomez-Garcia, M.A., V. Pitchon, and A. Kiennemann, *Pollution by nitrogen oxides: an approach to NO$_x$ abatement by using sorbing catalytic materials.* Environment International, 2005 **31** (3): p. 445-467.
29. Furuya, T., et al., *Development of a hybrid catalytic combustor for a 1300°C class gas turbine.* Catalysis Today, 1995 **26** (3-4): p. 345-350.
30. Sadamori, H., *Application concepts and evaluation of small-scale catalytic combustors for natural gas.* Catalysis Today, 1999 **47** (1-4): p. 325-338.
31. Furuya, T., et al., *nitrogen oxides decreasing combustion method*, in *United States Patent*. 1988, Kabushiki Kaisha Toshiba, Japan: US. p. 20.
32. Eguchi, K. and H. Arai, *Recent advances in high temperature catalytic combustion.* Catalysis Today, 1996 **29** (1-4): p. 379-386.
33. Sadamori, H., T. Tanioka, and T. Matsuhisa, *Development of a high-temperature combustion catalyst system and prototype catalytic combustor turbine test results.* Catalysis Today, 1995 **26** (3-4): p. 337-344.
34. Carroni, R., V. Schmidt, and T. Griffin, *Catalytic combustion for power generation.* Catalysis Today, 2002 **75** (1-4): p. 287-295.
35. Forzatti, P., *Status and perspectives of catalytic combustion for gas turbines.* Catalysis Today, 2003 **83** (1-4): p. 3-18.
36. Radojevic, M., *Reduction of nitrogen oxides in flue gases.* Environmental Pollution Nitrogen, 1998 **102** (1, Supplement 1): p. 685-689.
37. Narula, C.K., et al., *Materials Issues Related to Catalysts for Treatment of Diesel Exhaust.* Int. Journal of Applied Ceramic Technology, 2005 **2** (6): p. 452-466.
38. Bosch, H. and F. Janssen, *Preface.* Catalysis Today, 1988 **2** (4): p. v.
39. Iwamoto, M. and H. Hamada, *Removal of nitrogen monoxide from exhaust gases through novel catalytic processes.* Catalysis Today, 1991 **10** (1): p. 57-71.
40. Kikuchi, E., et al., *Selective Reduction of Nitric Oxide with Methane on Gallium and Indium Containing H-ZSM-5 Catalysts: Formation of Active Sites by Solid-State Ion Exchange.* Journal of Catalysis, 1996 **161** (1): p. 465-470.
41. Li, Y. and J.N. Armor, *Catalytic reduction of nitrogen oxides with methane in the presence of excess oxygen.* Applied Catalysis B: Environmental, 1992 **1** (4): p. L31-L40.
42. Ferri, D., et al., *NO reduction by H$_2$ over perovskite-like mixed oxides.* Applied Catalysis B: Environmental, 1998 **16** (4): p. 339-345.
43. Takahashi, N., et al., *The new concept 3-way catalyst for automotive lean-burn engine: NO$_x$ storage and reduction catalyst.* Catalysis Today, 1996 **27** (1-2): p. 63-69.
44. Mulla, S.S., et al., *Regeneration mechanism of Pt/BaO/Al$_2$O$_3$ lean NO$_x$ trap catalyst with H$_2$.* Catalysis Today, 2008 **15** (1-2): p. 136-145.
45. Breen, J.P., et. al., *Sulphur-Tolerant NO$_x$ Storage Traps: An Infrared and Thermodynamic Study of the Reactions of Alkali and Alkaline-Earth Metal Sulfates.* Catalysis Letters, 2002 **80** (3): p. 123-128.
46. Courson, C., et al., *Experimental study of the SO$_2$ removal over a NO$_x$ trap catalyst.* Catalysis Communications, 2002 **3** (10): p. 471-477.

47. Rappe, K.G., et al., *Combination of low and high temperature catalytic materials to obtain broad temperature coverage for plasma-facilitated NO_x reduction.* Catalysis Today, 2004 **89** (1-2): p. 143-150.
48. Li, J., et al., *A comparison study on non-thermal plasma-assisted catalytic reduction of NO by C_3H_6 at low temperatures between Ag/USY and Ag/Al_2O_3 catalysts.* Catalysis Today, 2007 **126** (3-4): p. 272-278.
49. Armor, J.N., *NO_x/hydrocarbon reactions over gallium loaded zeolites: A review.* Catalysis Today, 1996 **31** (3-4): p. 191-198.
50. Iwamoto, M. and H. Yahiro, *Novel catalytic decomposition and reduction of NO.* Catalysis Today, 1994 **22** (1): p. 5-18.
51. Desai, A.J., et al., *CoZSM-5: Why This Catalyst Selectively Reduces NO_x with Methane.* Journal of Catalysis, 1999 **184** (2): p. 396-405.
52. Armor, J.N., *Catalytic reduction of nitrogen oxides with methane in the presence of excess oxygen: A review.* Catalysis Today, 1995 **26** (2): p. 147-158.
53. Krishna, K. and M. Makkee, *Coke formation over zeolites and CeO_2-zeolites and its influence on selective catalytic reduction of NO_x.* Applied Catalysis B: Environmental, 2005 **59** (1-2): p. 35-44.
54. Praserthdam, P. and Ayutthaya, S.I.n., *Roles of NO and O_2 on coke deposition and removal over Cu-ZSM-5.* Catalysis Today, 2004 **97** (2-3): p. 137-143.
55. Capek, L., et al., *Cu-ZSM-5 zeolite highly active in reduction of NO with decane under water vapor presence: Comparison of decane, propane and propene by in situ FTIR.* Applied Catalysis B: Environmental, 2005 **60** (3-4): p. 201-210.
56. Jen, H.-W., *Study of nitric oxide reduction over silver/alumina catalysts under lean conditions: Effects of reaction conditions and support.* Catalysis Today, 1998 **42** (1-2): p. 37-44.
57. She, X. and M. Flytzani-Stephanopoulos, *The role of AgOAl species in silver-alumina catalysts for the selective catalytic reduction of NO_x with methane.* Journal of Catalysis, 2006 **237** (1): p. 79-93.
58. Miyadera, T., *Alumina-supported silver catalysts for the selective reduction of nitric oxide with propene and oxygen-containing organic compounds.* Applied Catalysis B: Environmental, 1993 **2** (2-3): p. 199-205.
59. Meunier, F.C., et al., *Mechanistic Aspects of the Selective Reduction of NO by Propene over Alumina and Silver-Alumina Catalysts.* Journal of Catalysis, 1999 **187** (2): p. 493-505.
60. Shimizu, K.I., A. Satsuma, and T. Hattori, *Catalytic performance of Ag-Al_2O_3 catalyst for the selective catalytic reduction of NO by higher hydrocarbons.* Applied Catalysis B: Environmental, 2000 **25** (4): p. 239-247.
61. Burch, R. and A. Ramli, *A kinetic investigation of the reduction of NO by CH_4 on silica and alumina-supported Pt catalysts.* Applied Catalysis B: Environmental, 1998 **15** (1-2): p. 63-73.
62. Hamada, H., *Selective reduction of NO by hydrocarbons and oxygenated hydrocarbons over metal oxide catalysts.* Catalysis Today, 1994 **22** (1): p. 21-40.
63. Tanaka, H. and M. Misono, *Advances in designing perovskite catalysts.* Current Opinion in Solid State and Materials Science, 2001 **5** (5): p. 381-387.
64. Dai, H., et al., *The relationship of structural defect-redox property-catalytic performance of perovskites and their related compounds for CO and NO_x removal.* Catalysis Today, 2004 **90** (3-4): p. 231-244.
65. Teraoka, Y., K. Kanada, and S. Kagawa, *Synthesis of La---K---Mn---O perovskite-type oxides and their catalytic property for simultaneous removal of NO_x and diesel soot particulates.* Applied Catalysis B: Environmental, 2001 **34** (1): p. 73-78.
66. Rozier, P., K. Jansson, and M. Nygren, *Investigation of structural and catalytic properties of $BaLa_4Cu_{5-y}Ru_yO_{13-\delta}$ with $0.0 \leq y \leq 3.0$ and of $LaBaCuRuO_6$.* Materials Research Bulletin, 2000 **35** (9): p. 1391-1400.

67. Dacquin, J.P., C. Dujardin, and P. Granger, *Catalytic decomposition of N_2O on supported Pd catalysts: Support and thermal ageing effects on the catalytic performances.* Catalysis Today, 2008 **137** (2-4): p. 390-396.
68. Dacquin, J.P., C. Dujardin, and P. Granger, *Surface reconstruction of supported Pd on $LaCoO_3$: Consequences on the catalytic properties in the decomposition of N_2O.* Journal of Catalysis, 2008 **253** (1): p. 37-49.
69. Ishihara, T., et al., *Direct decomposition of NO into N_2 and O_2 over $La(Ba)Mn(In)O_3$ perovskite oxide.* Journal of Catalysis, 2003 **220** (1): p. 104-114.
70. Marnellos, G.E., E.A. Efthimiadis, and I.A. Vasalos, *Effect of SO_2 and H_2O on the N_2O decomposition in the presence of O_2 over Ru/Al_2O_3.* Applied Catalysis B: Environmental, 2003 **46** (3): p. 523-539.
71. Tanaka, H., et al., *$LaFePdO_3$ perovskite automotive catalyst having a self-regenerative function.* Journal of Alloys and Compounds, 2006 **408-412**: p. 1071-1077.
72. Tanaka, H., et al., *The intelligent catalyst having the self-regenerative function of Pd, Rh and Pt for automotive emissions control.* Catalysis Today, 2006 **117** (1-3): p. 321-328.
73. Nishihata, Y., et al., *Self-regeneration of palladium-perovskite catalysts in modern automobiles.* Journal of Physics and Chemistry of Solids, 2005 **66** (2-4): p. 274-282.
74. Uenishi, M., et al., *Redox behavior of palladium at start-up in the Perovskite-type $LaFePdO_x$ automotive catalysts showing a self-regenerative function.* Applied Catalysis B: Environmental, 2005 **57** (4): p. 267-273.
75. Engelmann-Pirez, M., P. Granger, and G. Leclercq, *Investigation of the catalytic performances of supported noble metal based catalysts in the $NO + H_2$ reaction under lean conditions.* Catalysis Today, 2005 **107-108**: p. 315-322.
76. Uenishi, M., et al., *The reducing capability of palladium segregated from perovskite-type $LaFePdO_x$ automotive catalysts.* Applied Catalysis A: General, 2005 **296** (1): p. 114-119.
77. Singh, U.G., et al., *A Pd-doped perovskite catalyst, $BaCe_{1-x}Pd_xO_{3-\delta}$, for CO oxidation.* Journal of Catalysis, 2007 **249** (2): p. 349-358.
78. Schulz, U. and M. Schmücker, *Microstructure of ZrO_2 thermal barrier coatings applied by EB-PVD.* Materials Science and Engineering A, 2000 **276** (1-2): p. 1-8.
79. Movchan, B.A. and A.V. Demchishin, *Study of the structure and properties of thick vacuum condensates of nickel, titanium, tungsten, aluminium oxide and zirconium oxide.* Fitz. Metal. Metalloved, 1969 **28**: p. 653-660.
80. Liu, Z., et al., *Study of $Ag/La_{0.6}Ce_{0.4}CoO_3$ catalysts for direct decomposition and reduction of nitrogen oxides with propene in the presence of oxygen.* Applied Catalysis B: Environmental, 2003 **44** (4): p. 355-370.
81. McDaniel, C.L. and S.J. Schneider, *Phase relations between palladium oxide and the rare earth sesquioxides in air.* J. Res. Nat. Bur. Stand., **72** A: 27-37 (Jan.-Feb. 1968). 1968.
82. Tanaka, H., N. Mizuno, and M. Misono, *Catalytic activity and structural stability of $La_{0.9}Ce_{0.1}Co_{1-x}Fe_xO_3$ perovskite catalysts for automotive emissions control.* Applied Catalysis A: General, 2003 **244** (2): p. 371-382.
83. Costa, C.N. and A.M. Efstathiou, *Low-temperature H_2-SCR of NO on a novel Pt/MgO-CeO_2 catalyst.* Applied Catalysis B: Environmental, 2007 **72** (3-4): p. 240-252.
84. Hori, C.E., et al., *Thermal stability of oxygen storage properties in a mixed CeO_2-ZrO_2 system.* Applied Catalysis B: Environmental, 1998 **16** (2): p. 105-117.
85. Pimentel, P.M., et al., *Pechini Synthesis and Microstructure of Nickel-Doped Copper Chromites.* Materials Research Bulletin, 2005 **8**: p. 221-224.
86. Briggs, D. and M.P. Seah *Practical Surface Analysis by Auger and X-ray Photoelectron Spectroscopy*, ed. J.W.S. Ltd. 1983. 533.
87. Schneider, J., *WYRIET for Rietveld Analysis, Institut für Kristallographie*. 1994, Universität München: München.

88. Stranzenbach, M., *Entwicklung von integrierbaren Impedanz- NO_x-Gassensoren für den Hochtemperatureinsatz in extremen Bedingungen, Dissertation.* 2008, Brandenburgische Technische Universität Cottbus: Cottbus. p. 131.
89. McReady, D.E. and J.J. Kingsley, *Powder Data for $LaCo_{0.4}Fe_{0.6}O_3$.* Powder Diffraction, 1994 **9** (2): p. 143-145.
90. Kim, Y.I., Jung, J.K., and Ryu, K.S., *Structural study of nano $BaTiO_3$ powder by Rietveld refinement.* Materials Research Bulletin, 2004 **39** (7-8): p. 1045-1053.
91. Mandal, T.K., *Characterization of tetragonal $BaTiO_3$ nanopowders prepared with a new soft chemistry route.* Materials Letters, 2007 **61** (3): p. 850-854.
92. Makshina, E.V., et al., *Characterization and catalytic properties of nanosized cobaltate particles prepared by in situ synthesis inside mesoporous molecular sieves.* Applied Catalysis A: General, 2006 **312**: p. 59-66.
93. Tang, Z., Z. Zhou, and Z. Zhang, *Experimental study on the mechanism of BaTiO3-based PTC-CO gas sensor.* Sensors and Actuators B: Chemical, 2003 **93** (1-3): p. 391-395.
94. Kumar, S., Raju, V.S., and Kutty, T.R.N., *Investigations on the chemical states of sintered barium titanate by X-ray photoelectron spectroscopy.* Applied Surface Science, 2003 **206** (1-4): p. 250-261.
95. Schulz, U., S.G. Terry, and C.G. Levi, *Microstructure and texture of EB-PVD TBCs grown under different rotation modes.* Materials Science and Engineering A, 2003 **360** (1-2): p. 319-329.
96. Graham, L.A., et al., *Greenhouse gas emissions from heavy-duty vehicles.* Atmospheric Environment, 2008 **42** (19): p. 4665-4681.
97. Zhou, K., et al., *Pd-containing perovskite-type oxides used for three-way catalysts.* Journal of Molecular Catalysis A: Chemical, 2002. **189**(2): p. 225-232.
98. Rodríguez G.C.M., *Meta-stability and microstructure of the $LaFe_{0.65}Co_{0.3}Pd_{0.05}O_3$ perovskite compound prepared by a modified citrate route.* Journal of European Ceramic Society, 2008 **28** (13): p. 2611-2616.
99. Royer, S., F. Bérubé, and S. Kaliaguine, *Effect of the synthesis conditions on the redox and catalytic properties in oxidation reactions of $LaCo_{1-x}Fe_xO_3$.* Applied Catalysis A: General, 2005 **282** (1-2): p. 273-284.
100. Twagirashema, I., et al., *An in situ study of the $NO + H_2 + O_2$ reaction on $Pd/LaCoO_3$ based catalysts.* Catalysis Today, 2007 **119** (1-4): p. 100-105.
101. Barrera, A., et al., *The role of lanthana loading on the catalytic properties of Pd/Al_2O_3-La_2O_3 in the NO reduction with H_2.* Applied Catalysis B: Environmental, 2005 **56** (4): p. 279-288.
102. Wagner, C.D., et al., *NIST X-ray Photoelectron Spectroscopy Database.* 2007, U.S. Department of Commerce: U.S.
103. Breen, J.P., et al., *A fast transient kinetic study of the effect of H_2 on the selective catalytic reduction of NO_x with octane using isotopically labelled 15NO.* Journal of Catalysis, 2007 **246** (1): p. 1-9.
104. Breen, J.P., et al., *An investigation of the thermal stability and sulphur tolerance of Ag/γ-Al_2O_3 catalysts for the SCR of NO_x with hydrocarbons and hydrogen.* Applied Catalysis B: Environmental, 2007 **70** (1-4): p. 36-44.
105. Satokawa, S., et al., *Promotion effect of hydrogen on lean NO_x reduction by hydrocarbons over Ag/Al_2O_3 catalyst.* Chemical Engineering Science, 2007 **62** (18-20): p. 5335-5337.
106. Macleod, N. and R.M. Lambert, *Lean NO_x reduction with $CO + H_2$ mixtures over Pt/Al_2O_3 and Pd/Al_2O_3 catalysts.* Applied Catalysis B: Environmental, 2002 **35** (4): p. 269-279.
107. Qi, G., R.T. Yang, and F.C. Rinaldi, *Selective catalytic reduction of nitric oxide with hydrogen over Pd-based catalysts.* Journal of Catalysis, 2006 **237** (2): p. 381-392.
108. Dhainaut, F., S. Pietrzyk, and P. Granger, *Kinetic investigation of the NO reduction by H_2 over noble metal based catalysts.* Catalysis Today, 2007 **119** (1-4): p. 94-99.

109. Nitadori, T. and M. Misono, *Catalytic properties of $La_{1-x}A'_xFeO_3$ (A' = Sr, Ce) and $La_{1-x}Ce_xCoO_3$*. Journal of Catalysis, 1985 **93** (2): p. 459-466.
110. Hodjati, S., et al., *Absorption/desorption of NO_x process on perovskites: performances to remove NO_x from a lean exhaust gas*. Applied Catalysis B: Environmental, 2000 **26** (1): p. 5-16.
111. Buciuman, F.-C., et al., *Catalytic properties of $La_{0.8}A_{0.2}MnO_3$ (A = Sr, Ba, K, Cs) and $LaMn_{0.8}B_{0.2}O_3$ (B = Ni, Zn, Cu) perovskites: 2. Reduction of nitrogen oxides in the presence of oxygen*. Applied Catalysis B: Environmental, 2001.**35** (2): p. 149-156.
112. Huuhtanen, M., et al., *In situ FTIR study on NO reduction by C_3H_6 over Pd-based catalysts*. Catalysis Today, 2002 **75** (1-4): p. 379-384.
113. Wada, K., et al., *Texture and microstructure of ZrO_2-4mol% Y_2O_3 layers obliquely deposited by EB-PVD*. Surface and Coatings Technology, 2006 **200** (8): p. 2725-2730.
114. Rentería, A.F., *A small -angle scattering analysis of the influence of manufacture and thermal induced morphological changes on the thermal conductivity of EB-PVD Thermal Barrier Coatings*, in *Fakultät für Georessourcen und Materialtechnik der Rheinisch-Westfälischen Technischen Hochschule Aachen*. 2007, RTWHA: Aachen. p. 142.
115. Renteria, A.F., et al., *Effect of morphology on thermal conductivity of EB-PVD PYSZ TBCs*. Surface and Coatings Technology, 2006 **201** (6): p. 2611-2620.
116. Uwe Schulz, *Phase Transformation in EB-PVD Ytria Partially Stabilized Zirconia Thermal Barrier Coatings during Annealing*. Journal of the American Ceramic Society, 2000 **83** (4): p. 904-910.
117. Frank, E., *Modellierung und Simulation der katalysierten Reduktion von NO_x mittels Propen in sauerstoffreichen Angasen an Wabenkatalysatoren*, in *Fakultät für Chemie*. 2002, TH Universität Karlsruhe: Karlsruhe. p. 151.

Appendix

Appendix A1: Rietveld refinement of the lanthanum based perovskite.

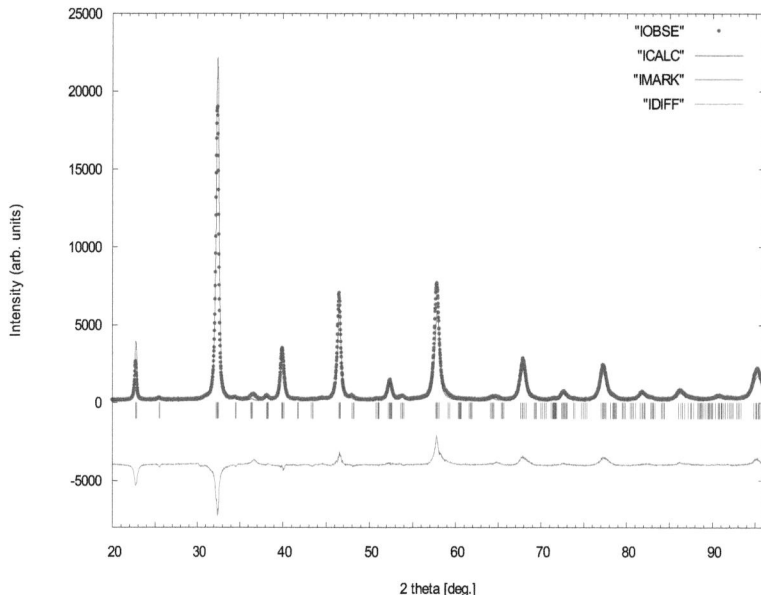

Figure A1.1 Rietveld refinement of LaFeCo$_{(0.35)}$ calcined in air at 700°C/3h (orthorhombic perovskite) employing a single phase model.

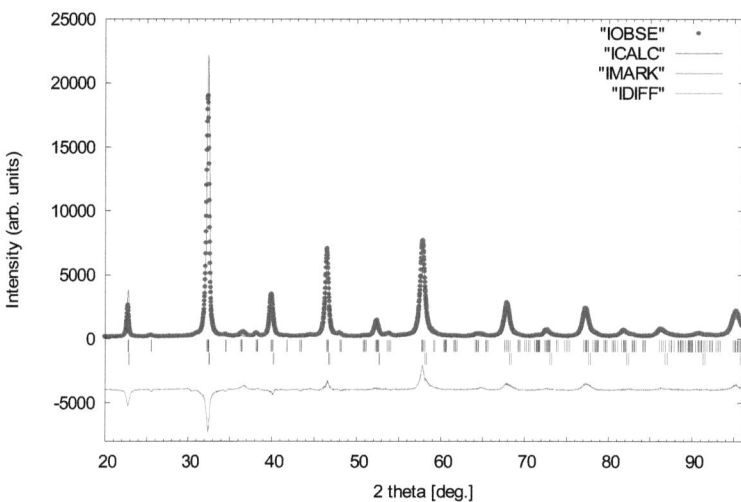

Figure A1.2 Rietveld refinement of LaFeCo$_{(0.35)}$ calcined in air at 700°C/3h employing a two phase model including orthorhombic and cubic perovskite phases.

Appendix

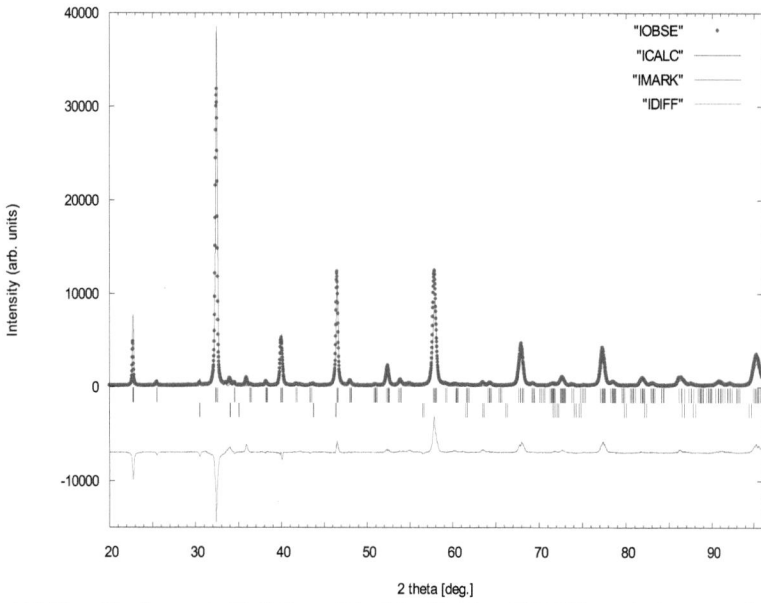

Figure A1.3 Rietveld refinement of $LaFeCo_{(0.3)}$-Pd calcined in air at 900°C/3h employing a two phase model including orthorhombic perovskite and tetragonal PdO.

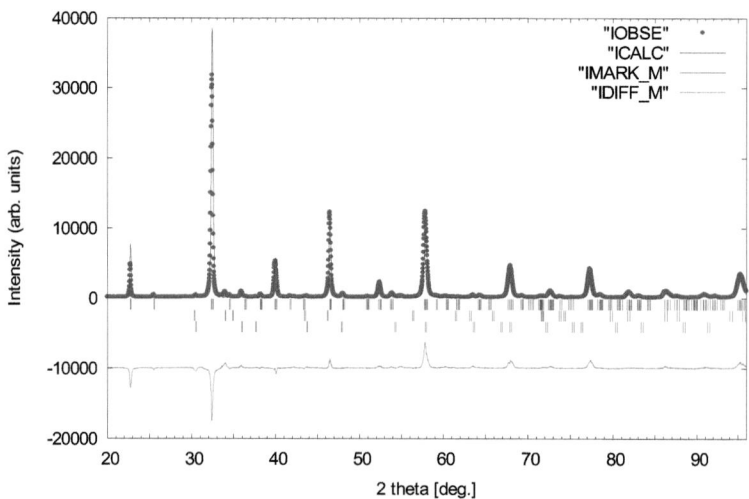

Figure A1.4 Rietveld refinement of $LaFeCo_{(0.3)}$-Pd calcined in air at 900°C/3h employing a three phase model including orthorhombic perovskite, cubic Co_3O_4 and tetragonal PdO.

Appendix A2: Activity test of the catalytic converter during C_3H_6-SCR of NO_x

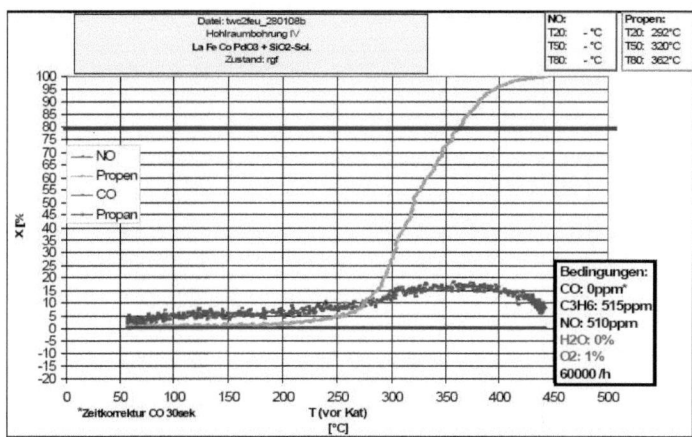

Figure A2.1 NO_x-conversion during the C_3H_6-SCR of NO_x over the catalytic converter $LaFeCo_{(0.3)}$-Pd-cordierite carried out at INTERKAT GmbH[®]. *Feed composition: 510 ppm NO, 515 ppm C_3H_6, 1 vol. % O_2, balance N_2, heating rate = $10°K.min^{-1}$ at SV = $60000\ h^{-1}$.*

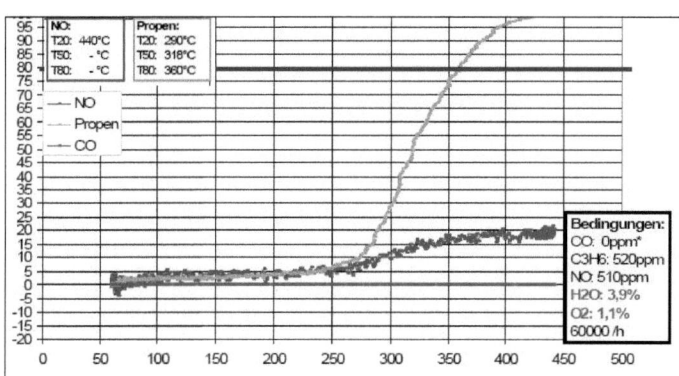

Figure A2.2 NO_x-conversion during the C_3H_6-SCR of NO_x over the catalytic converter $LaFeCo_{(0.3)}$-Pd-cordierite carried out at INTERKAT GmbH[®]. *Feed composition: 510 ppm NO, 520 ppm C_3H_6, 1.1 vol. % O_2, 3.9 vol. % H_2O_{vapour}, balance N_2, heating rate = $10°K.min^{-1}$ at SV = $60000\ h^{-1}$. Already reported in section 5.3.5.*

Appendix

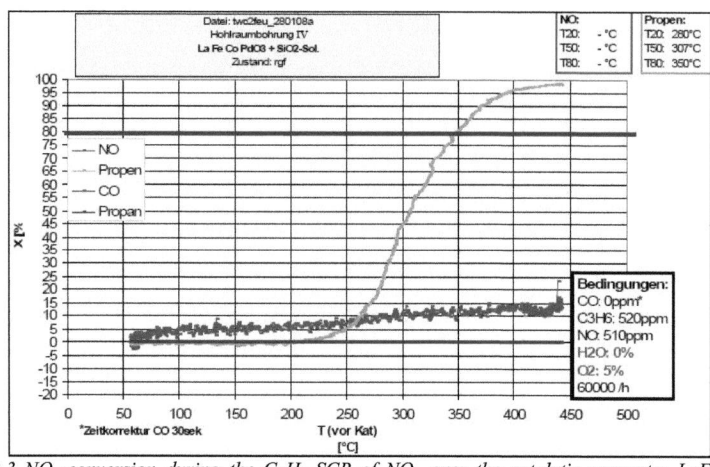

Figure A2.3 NO_x-conversion during the C_3H_6-SCR of NO_x over the catalytic converter $LaFeCo_{(0.3)}$-Pd-cordierite carried out at INTERKAT GmbH®. *Feed composition: 510 ppm NO, 520 ppm C_3H_6, 5 vol. % O_2, balance N_2, heating rate = $10°K.min^{-1}$ at SV = 60000 h^{-1}.*

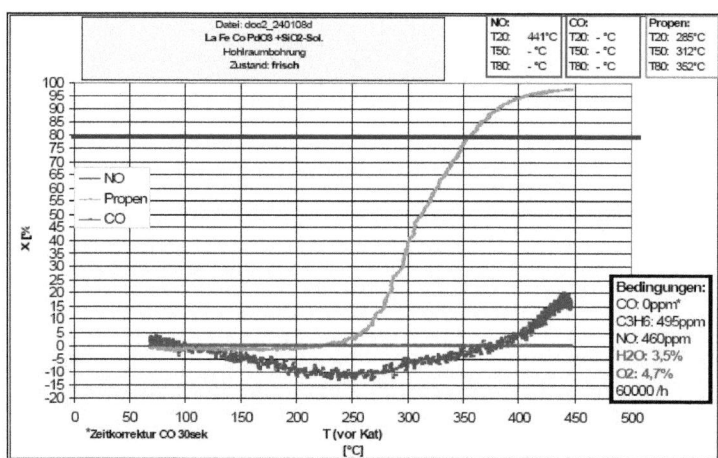

Figure A2.4 NO_x-conversion during the C_3H_6-SCR of NO_x over the catalytic converter $LaFeCo_{(0.3)}$-Pd-cordierite carried out at INTERKAT GmbH®. *Feed composition: 460 ppm NO, 495 ppm C_3H_6, 4.7 vol. % O_2, 3.5 vol. % H_2O_{vapour}, balance N_2, heating rate = $10°K.min^{-1}$ at SV = 60000 h^{-1}.*

Appendix

Appendix A3: Calibration curves for N_2O measurement under the selected reaction conditions

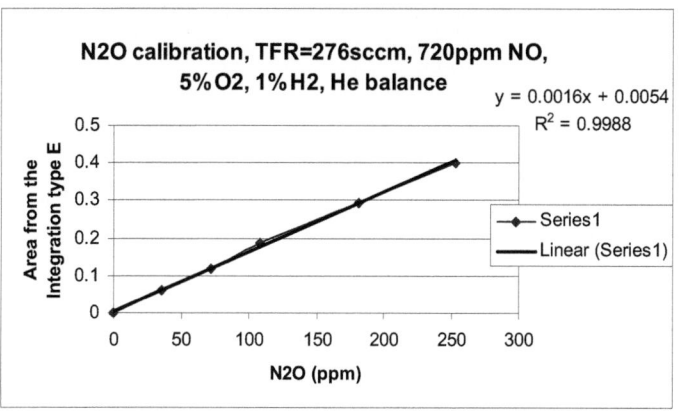

Figure A3.1 N_2O calibration in the mixture 720 ppm NO, 5 vol. % O_2, 1 vol. % H_2 and He as balance, TFR = 276 ml.min^{-1}.

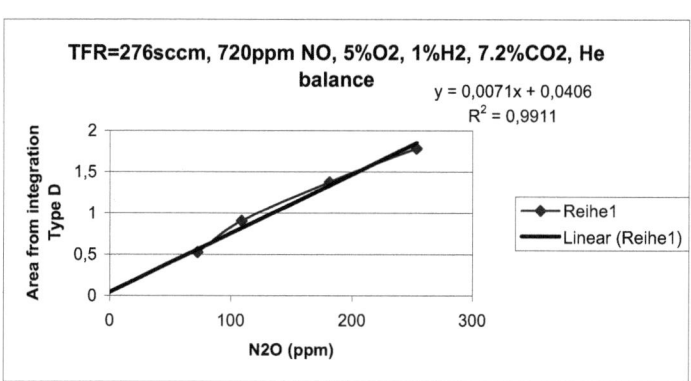

Figure A3.2 N_2O calibration in the mixture 720 ppm NO, 5 vol. % O_2, 1 vol. % H_2, 7.2 vol. % CO_2 and He as balance, TFR = 276 ml.min^{-1}.

Appendix A4: Catalysts formulations

Catalyst's formulations synthesized and studied in this study.

Citrate route	Sample Code
$LaFe_{0.475}Co_{0.475}\mathbf{Pd_{0.05}}O_3$	$LaFeCo_{(0.475)}$-Pd
$LaFe_{0.475}Co_{0.475}\mathbf{Rh_{0.05}}O_3$	$LaFeCo_{(0.475)}$-Rh
$LaFe_{0.55}Co_{0.4}\mathbf{Pd_{0.05}}O_3$	$LaFeCo_{(0.4)}$-Pd
$La_{0.95}Ce_{0.05}Fe_{0.65}Co_{0.3}\mathbf{Pd_{0.05}}O_3$	$LaCe_{(0.05)}FeCo_{(0.3)}$-Pd
$La_{0.95}Ce_{0.05}Fe_{0.95}\mathbf{Pd_{0.05}}O_3$	$LaCe_{(0.05)}$Fe-Pd
$La_{0.6}Ce_{0.4}Fe_{0.95}\mathbf{Pd_{0.05}}O_3$	$LaCe_{(0.4)}$Fe-Pd
$LaFe_{0.65}Co_{0.3}\mathbf{Pd_{0.05}}O_3$ *	$LaFeCo_{(0.3)}$-Pd
$\mathbf{Pd}$-$LaFe_{0.65}Co_{0.35}O_3$ *	Pd- $LaFeCo_{(0.3)}$
$LaFe_{0.65}Co_{0.35}O_3$ *	$LaFeCo_{(0.35)}$
Co-precipitation method	
$BaTi_{0.95}\mathbf{Pd_{0.05}}O_3$ *	BaTi-Pd
$\mathbf{Pd}$-$BaTiO_3$ *	Pd-BaTi
$BaTiO_3$ *	BaTi

* Catalysts compositions subjected to main investigations

I want morebooks!

Buy your books fast and straightforward online - at one of the world's fastest growing online book stores! Environmentally sound due to Print-on-Demand technologies.

Buy your books online at

www.get-morebooks.com

Kaufen Sie Ihre Bücher schnell und unkompliziert online – auf einer der am schnellsten wachsenden Buchhandelsplattformen weltweit!
Dank Print-On-Demand umwelt- und ressourcenschonend produziert.

Bücher schneller online kaufen
www.morebooks.de

OmniScriptum Marketing DEU GmbH
Heinrich-Böcking-Str. 6-8
D - 66121 Saarbrücken
Telefax: +49 681 93 81 567-9

info@omniscriptum.com
www.omniscriptum.com

Printed by Books on Demand GmbH, Norderstedt / Germany